ADVANCED WOODWORKING PROJECTS

ADVANCED WOODWORKING PROJECTS

ROBERT LENTO

PRENTICE-HALL, INC., Englewood Cliffs, N.J. 07632

Library of Congress Cataloging-in-Publication Data

Lento, Robert
  Advanced woodworking projects.

    1. Woodwork.   2. Furniture making.   I. Title.
  II. Title: Advanced woodworking projects.
  TT180.L427 1987   684.1′042   86-20492
  ISBN   0-13-011834-6

Editorial/production supervision and
  interior design: Ellen Denning
Cover design: 20/20 Services, Inc.
Manufacturing buyer: John Hall
Page layout: Steve Frimm

Printed in the United States of America

10 9 8 7 6 5 4 3 2 1

ISBN   0-13-011834-6   025

PRENTICE-HALL INTERNATIONAL (UK) LIMITED, *London*
PRENTICE-HALL OF AUSTRALIA PTY. LIMITED, *Sydney*
PRENTICE-HALL CANADA INC., *Toronto*
PRENTICE-HALL HISPANOAMERICANA, S.A., *Mexico*
PRENTICE-HALL OF INDIA PRIVATE LIMITED, *New Delhi*
PRENTICE-HALL OF JAPAN, INC., *Tokyo*
PRENTICE-HALL OF SOUTHEAST ASIA PTE. LTD., *Singapore*
EDITORA PRENTICE-HALL DO BRASIL, LTDA., *Rio de Janeiro*

This book is dedicated to the memory of Louis and Mary Lento
and Eugene L. Wurmser

Contents

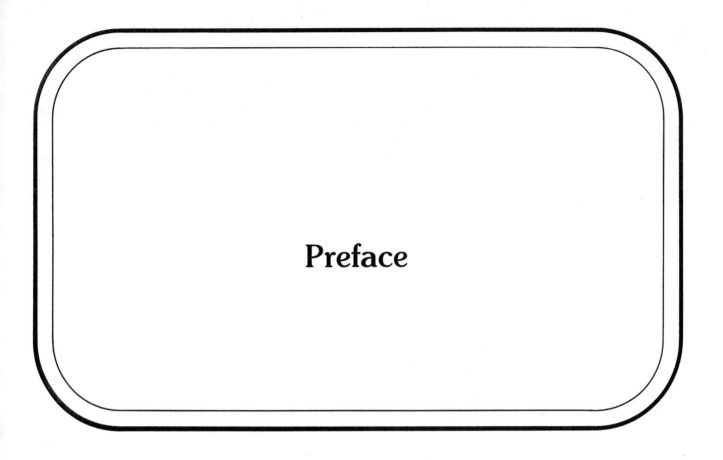

Preface

As a project book this book is probably not typical. Whereas most project books focus on a single furniture style, this book does not. In fact, it not only provides plans for furniture pieces of different styles, it also includes some non-furniture pieces. Most of the projects in the book require at least a basic knowledge of woodworking. The lower-numbered pieces are the simplest and in general require the least amount of experience and skill. In a similar way the amount of equipment required increases as the project number goes up. The early pieces can easily be made with hand tools. The later ones can also be made with hand tools, but power equipment will make the job much more feasible.

Among the furniture pieces in the collection will be found a variety of styles, ranging from period pieces to contemporary designs. Several highly functional projects, such as the drafting table and valet chair, are presented.

All the projects are described visually in two or three views with important details given in extra views and exploded views. In addition, all the pieces are represented pictorially in perspective drawings. Some of the exploded views are isometric drawings, which are easier for the inexperienced eye to interpret. A bill of materials is provided for each project. This list describes the dimensions of each part and the number of identical pieces required. Specific construction notes give detailed information related to constructing individual parts.

In addition to the drawings, each piece is described verbally. General directions for fabricating the project are provided. This description supplements the drawings and the construction notes given in the bill of materials.

Figures 1 through 8 are from *Woodworking: Tools, Fabrication, Design and Manufacturing* by Robert Lento, © 1979. Reprinted by permission of Prentice-Hall, Inc., Englewood Cliffs, N.J.

Most of the project designs can be altered to conform to individual tastes and use requirements. Dowel joints or plugged screws can often be used to replace mortise-and-tenon joints. Solid-wood components can be replaced with hardwood-veneered plywood. Square legs can be replaced by turned legs. Wood finishes are left to the craftsman's discretion.

ROBERT LENTO

ADVANCED WOODWORKING PROJECTS

INTRODUCTION

General Project-Making Sequence

For the craftsmen who wish to follow a specific procedure in the fabrication of their pieces, the following procedure is recommended:

1. Obtain, revise, or design the project to be constructed.

2. If no drawing of the project is available, develop one from the information on hand.

3. If no bill of materials is available, make one up from the drawing.

4. Select and rough-cut lumber to oversized dimensions. An inch of extra length and half an inch of extra width are usually adequate for processing the stock to finished dimensions.

Most of the projects in this book utilize standard nominally dimensioned lumber, such as 2 × 4s or 1 × 6s. This nominal material is approximately 1½ in. thick and 3½ or 5½ in. wide, respectively. When selecting material, avoid warped or heavily knotted pieces.

5. After the pieces on the materials list have been cut to oversized width and length, they must be *squared up*. This process results in each square or rectangular piece being reduced to the finished dimensions given on the bill of materials. Additionally, all adjacent surfaces are made square, or at a 90° angle, to each other. In other words, all surfaces that meet at a corner form a 90° angle.

This is done by selecting one flat *face* as the reference or *working face*. Then one *edge* is made *smooth*, *straight*, and *square* to that face. This is the *working edge.* The next step is to cut one of the ends smooth, straight, and square to the working face and the working edge. This is the *working end.* Now the stock is cut to final length with the second end made smooth, straight, and square to the *working face* and *working edge.* The last step is to cut the piece to final width, with the second edge being made parallel to the *working edge* and square to the *working face.*

6. Lay out and fabricate all the project joints. If there are identical joints, they can be made with a single setup for each part of the joint. For example, if a dado (groove across the grain) is required, a table saw or router can be set up and all the pieces requiring the dado can be cut one after the other.

7. Curved or tapered cuts can be made at this point. Making these cuts earlier would have interfered with the layout and cutting of the joints.

8. Where possible, holes and recesses for hardware should be made now. In many situations these operations must await assembly of the project.

9. Preliminary sanding through medium or No. 150 grit is carried out next.

10. A trial assembly is made. This assembly is made without glue. This step helps to eliminate clamping problems. This is also the time to make any final fitting adjustments.

11. Apply glue, assemble, and clamp the project and allow the glue to cure, usually overnight.

12. Remove the clamps and scrape away the excess glue. If corners are to be rounded or otherwise shaped, this is usually the best time to do that. Complete final sanding through extra fine or No. 220 grit paper is done at this stage.

13. After removing dust with a tack rag, the finishing process can be started.

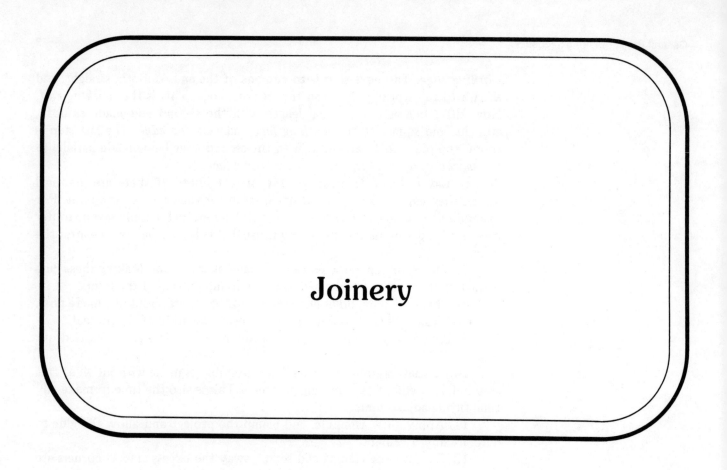

Joinery

The various projects in this book require a variety of joints. Some of the more frequently required joints are described in detail here. The specific dimensions for each joint are determined by checking the project drawing and individual part dimensions.

BUTT JOINTS   The butt joint is the simplest joint to make. The location of the pieces with respect to each other is determined by layout and clamping. Since pieces can shift before hardware is installed or glue sets, butt joints have to be checked several times during the securing process. Alignment lines on at least two surfaces can make such checking easier. The inclusion of nails, screws, or dowels in addition to glue makes butt joints more secure (Figures 1 and 2).

EDGE-TO-EDGE BUTT JOINT   The edge-to-edge butt joint is often used to join several narrow pieces together to form a single wider piece. Note that the *heartwood* is reversed in alternate pieces (Figure 3). This helps to limit cup warping. The diameter of the dowels used is determined by selecting dowels that are equal to one-third or one-half of the thickness of the pieces to be joined.

Dowel holes are bored on the centerline of the edges of each piece. Note the spacing of the holes [Figure 3(a)].

Figure 1 Fabricating a nailed butt joint.

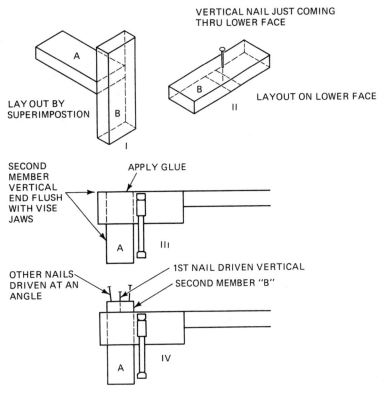

LAY OUT BY
SUPERIMPOSTION

VERTICAL NAIL JUST COMING
THRU LOWER FACE

LAYOUT ON LOWER FACE

SECOND
MEMBER
VERTICAL
END FLUSH
WITH VISE
JAWS

APPLY GLUE

OTHER NAILS
DRIVEN AT AN
ANGLE

1ST NAIL DRIVEN VERTICAL

SECOND MEMBER "B"

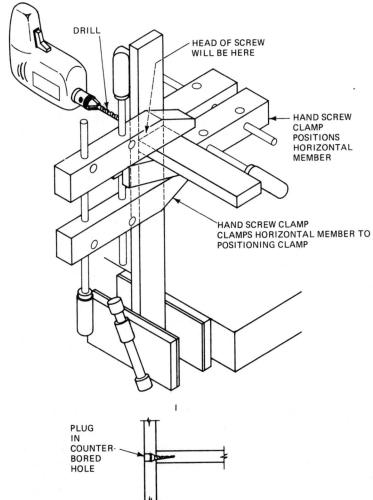

DRILL

HEAD OF SCREW
WILL BE HERE

HAND SCREW
CLAMP
POSITIONS
HORIZONTAL
MEMBER

HAND SCREW CLAMP
CLAMPS HORIZONTAL MEMBER TO
POSITIONING CLAMP

PLUG
IN
COUNTER-
BORED
HOLE

Figure 2 Fabricating a screwed or through dowel butt joint.

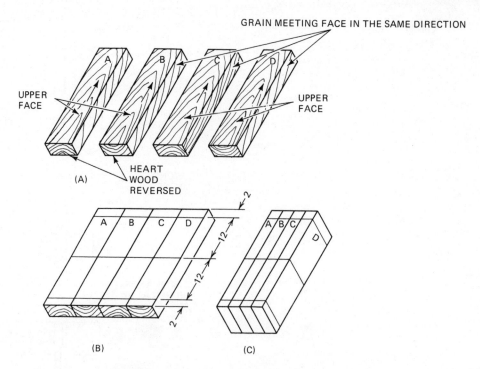

Figure 3   Fabricating an edge-to-edge dowel joint.

LAPPED JOINTS   Lapped joints are commonly used to join pieces meeting face up and at right angles to each other (Figure 4). Half the thickness of each piece is cut away so that when the pieces are mated, they add together to maintain the original thickness of each piece. The cutout areas are usually sawed in several places and then chiseled out. If power tools are available, a power router using a T-square guide can be used to remove the necessary material. A table saw fitted with dado cutters can be used to cut the recesses as the pieces are moved over the blade while being supported by a miter gauge.

DADO JOINTS   A dado joint is made by cutting a recess, usually halfway through the piece, into which the mating piece fits snugly (Figure 5). A variation of this joint, the rabbet dado, has a dado cut halfway through one piece into which a rabbet, a step cut in the end of the piece, fits. These joints are widely used in bookcase and drawer construction.

Dadoes can be made in much the same way that half-lapped joints are made. Dadoes can be cut with a backsaw and then chiseled to final size. The dado, a groove across the grain, is usually made slightly smaller than the mating piece. The mating piece is then scraped until it fits the dado snugly. A power router or table saw fitted with a dado cutter can also be used to cut the dadoes.

DOVETAIL JOINTS   Dovetail joints are among the most secure joints because their parts interlock. These joints find their most common application in drawer construction. Since drawer fronts are subjected to considerable pounding, the dovetail is the ideal joint to use in connecting drawer fronts to drawer sides.

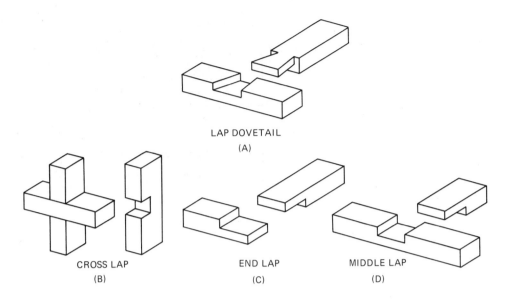

Figure 4 Lapped joints.

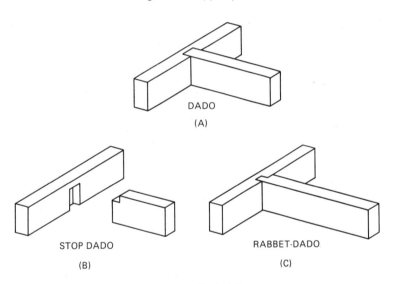

Figure 5 Dado joints.

These rather complex joints can be made most easily with a power router and a dovetail fixture designed to work with a particular machine. This combination of tools cuts matching pins and sockets in a single operation.

Handmade dovetails are difficult to make, but with practice they can be cut successfully. Figure 6 shows the layout procedure for making identical multiple dovetails. Many contemporary craftsmen make dovetail joints with a relatively small number of dovetails. This makes the job easier. In reality there are no hard-and-fast rules for dovetail joinery. As long as the slope of the individual dovetails is kept between 1 to 6 and 1 to 8, the joint should function successfully.

After the layout of the dovetail pins is completed on one piece, a series of fine saw cuts are made in the scrap areas of the joint parallel to the slanting layout lines. A suitable chisel is used to remove the excess material

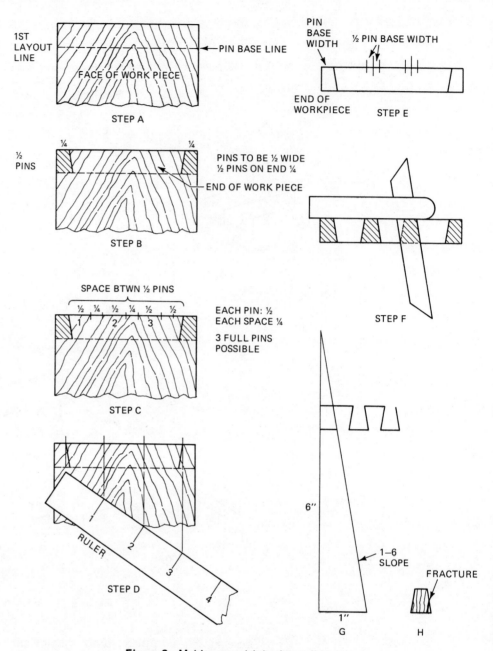

Figure 6 Making a multiple-dovetail layout.

between the saw cuts. A wider chisel is used to pare away the remaining wood until the layout lines on the pins are reached. The completed pins are superimposed on the second piece and traced. This provides a layout of the sockets to be cut. The sockets are cut in much the same way that the pins were. Final fitting is accomplished by blackening the pins with a pencil and then mating the two pieces. The pencil lead will indicate high areas on the second piece.

The joint design should allow for about 1/16 in. of extra wood on the pin ends. After the joint is assembled, this extra wood is sanded away, leaving the ends of the pins flush with the face of the piece surrounding them.

MORTISE-AND-TENON JOINTS

Mortise-and-tenon joints consist of a tongue or tenon which projects from one piece, fitting into a matching mortise, a rectangular recess in the second piece. These joints are very strong because they provide a large amount of surface area for glue. In addition, very little end grain is involved in the glued area. End grain does not glue up very effectively because it absorbs much of the glue. Mortise-and-tenon joints are used in chair and table construction, where strength is essential (Figure 7).

Tenon thickness is between one-third and one-half of the thickness of the piece on which it is cut. The maximum width of the tenon is usually limited to five times its thickness. Therefore, a tenon cut on a piece of wood ³/₄ in. thick might be ³/₈ in. thick and slightly less than 2 in. wide. The length of the tenon is usually limited to two-thirds the width or thickness of the piece into which it fits. At least ½ in. of material should be left between a mortise and the end of the piece into which it is cut.

Figure 7 Mortise-and-tenon joints.

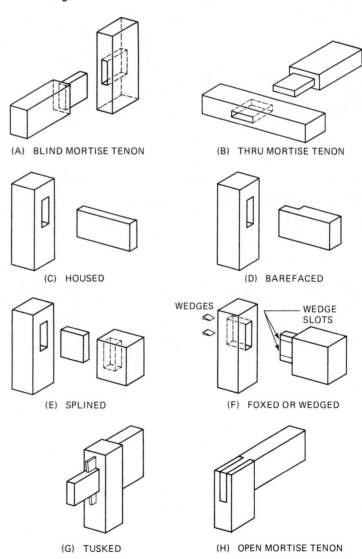

(A) BLIND MORTISE TENON   (B) THRU MORTISE TENON

(C) HOUSED   (D) BAREFACED

(E) SPLINED   (F) FOXED OR WEDGED

(G) TUSKED   (H) OPEN MORTISE TENON

In laying out this joint, the mortises are usually done first. Then the mortise layout is transferred to the piece to be tenoned, with a try square (Figure 8). Mortises are made first. This is accomplished by boring a series of holes along the centerline of the mortise layout. The bit used is equal in diameter to the width of the mortise. Each hole is bored to the same depth. A bit gauge can be used to limit the depth of each hole. The remaining excess wood around the holes is removed with a chisel. The tenon is cut to fit the existing mortise. It can be cut using the methods described in the section on lapped joints.

Figure 8 Mortise-and-tenon layout and fabrication.

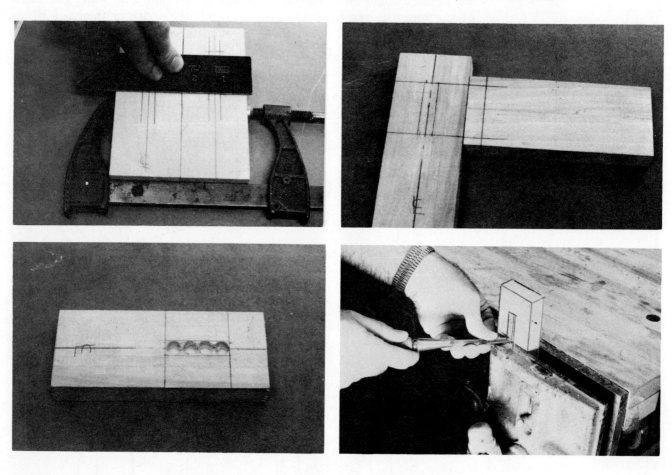

WOODWORKING
PROJECTS

PROJECT
1

Desktop Bookshelf

This useful project will serve as a bookshelf on a desk or table. Gravity is used to hold the books in place against the end of the piece.

The two largest parts of the project, the base and back, can be cut from a single piece of ½-in.-thick walnut 5½ in. wide and 28 in. long. The base piece, C, is squared up, and its front edge is planed down to ¼ in. thickness. This is accomplished by planing the taper shown in the end view of the drawing.

The back piece, D, is laid out, and then the slanting edge is ripped and planed to finished size. The upper corners of the back are rounded off to a ½-in. radius.

If power tools are used, the bevel on the base can be cut on a table saw. In this operation the blade is tilted to the angle formed between the tapered surface and the face of the piece. The piece is positioned with its face against the fence of the table saw and the taper or bevel is cut. The rough bevel is smoothed with a plane.

The end of the bookshelf is laid out and cut to its triangular shape. This can be done with a backsaw and plane or it can be rough-cut on a bandsaw and then sanded to final size and shape on a stationary belt sander. If a belt sander is not available, a block plane can be used to accomplish this.

The end piece is fastened to the base and back with through dowels. Quarter-inch-diameter dowels are appropriate. The holes through the base and back are located and bored. Then the end is clamped into position on the base while the holes are bored into the vertical edge of the end piece, using the holes in the back as a template. The process is reversed for the holes on the bottom of the end.

The small foot, A, and both parts of the large foot, B and B′, are right-angle triangles. These can be made by ripping a piece of ½-in. stock 2 in. wide and about 15 in. long. A miter box is used to cut three pieces. One, part B, is 5½ in. long; the second, part B′, is 4 in. long; and the third, part A, is 4¼ in. long. The triangle required is laid out on each of these pieces using the drawing as a guide. The pieces are then cut slightly oversized and planed to final size.

Parts B and B′ are half-lapped as shown on the drawing. The completed feet are fastened to the project base with through ¼-in-diameter dowels. If desired, the lower edges of the feet can be rounded off.

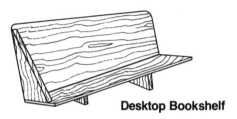

Desktop Bookshelf

Letter	Description	Number of pieces	Thick-ness	Width	Length
A	Small foot	1	1/2	1-1/4	4
B′	Back of large foot	1	1/2	1-3/4	5-5/16
B	Large–foot base	1	1/2	1-3/4	5-1/2
C	Base	1	1/2	5-1/2	14
D	Back	1	1/2	5-1/2	14
E	End	1	1/2	4	5

Notes
1. Material: walnut.
2. A and B fastened to C with through dowels.
3. E fastened to C and D with through dowels.

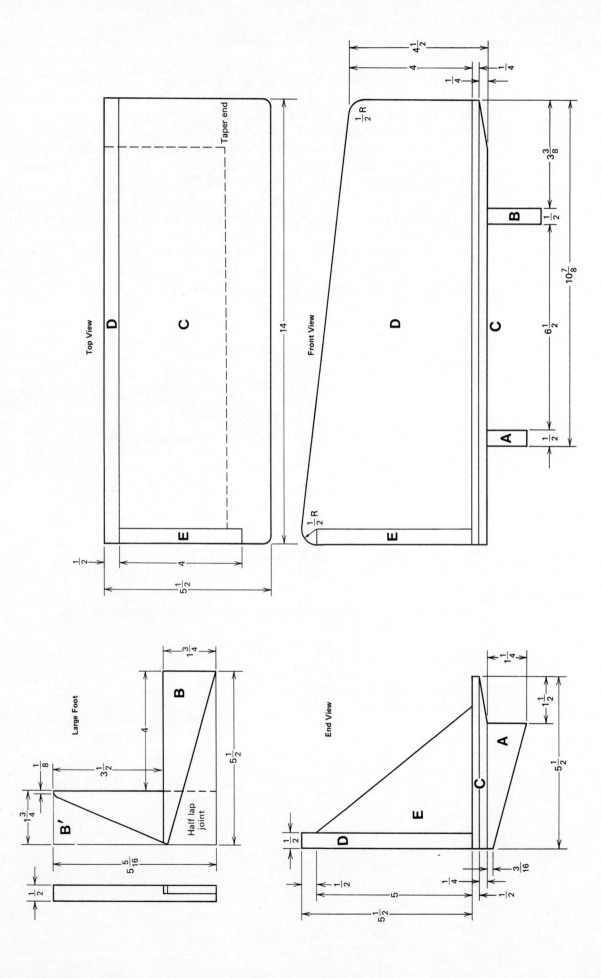

Top View

Front View

Large Foot

End View

Taper end

Half lap joint

14

PROJECT
2

Colonial Fishtail Pipe Box

Pipe boxes of this type were used in colonial times to store long-stemmed clay pipes. A new smoker would break off an inch or two of the stem and smoke the pipe. The drawer was used to hold tobacco. These pipes were often found in taverns or other public places.

This project is a fairly simple one to make. The first step is to square all the project's parts to finished dimensions. The ends of the back, front, and sides do not have to be squared up too carefully because they are going to be cut to curved shapes.

Cardboard templates are made to match the curved shapes which are to be cut on the ends of the back, front, and sides. This is accomplished by drawing a grid with 1-in. spaces on the cardboard which will be used to make the templates. This grid matches the grids that are drawn over the curved lines on the drawing. The grid lines, horizontal and vertical, on the drawing and on the cardboard are identified in the same way, using numbers on the horizontal lines and letters on the vertical. The points of intersection of the curved lines on the grid lines are reproduced on the grid lines on the cardboard. These points of intersection are connected by a smoothly flowing curved line and the cardboard is cut to shape. This results in a template that can be used to reproduce the drawing curves full size on the wooden parts.

The sides, E, can be fastened together face to face with double-sided carpet tape and cut as a single piece. This will result in two identical sides. The curves can be cut with a coping saw, scroll saw, or bandsaw.

The back, D, and the front, C, are laid out with half templates which are flipped over to make the second half of the layout. This will ensure a symmetrical pattern.

The bottom of the box, A, has its front edge and ends shaped as shown. The rear edge is a left square. Note that the sides of the back, D, are cut out to receive the side pieces, E. The edges of the front, E, are rabbeted (have a step cut out) to fit against the sides, E.

The major parts of the box are glued together before the drawer is made. The drawer is constructed to fit the existing opening. The drawer's construction can be simplified for ease of construction or it can be made nonfunctional by attaching (gluing) the drawer front in place. All corners should be rounded to approximately a ⅛-in. radius.

Colonial Pipe Box

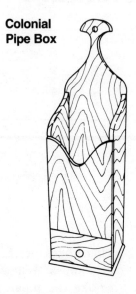

Letter	Description	Number of pieces	Thick-ness	Width	Length
A	Base	1	3/8	4-3/8	5-1/2
B	Drawer front	1	3/8	2-1/2	4-3/4
C	Box front	1	3/8	4-3/4	7-1/8
D	Back	1	3/8	4-3/4	17-3/4
E	Side	2	3/8	3-11/16	12
F	Drawer side	2	3/8	2-1/8	3-1/16
G	Drawer back	1	1/4	2-1/8	4
H	Drawer bottom	1	1/8	3-5/16	4
I	Box bottom	1	3/8	3-1/8	4

Notes
1. Material: white pine.
2. Drawer pull centered on drawer front.
3. Assemble sides to back; install box bottom, I, then attach front, C, then base, A.
4. Round off all outside corners 1/8 in. in radius.

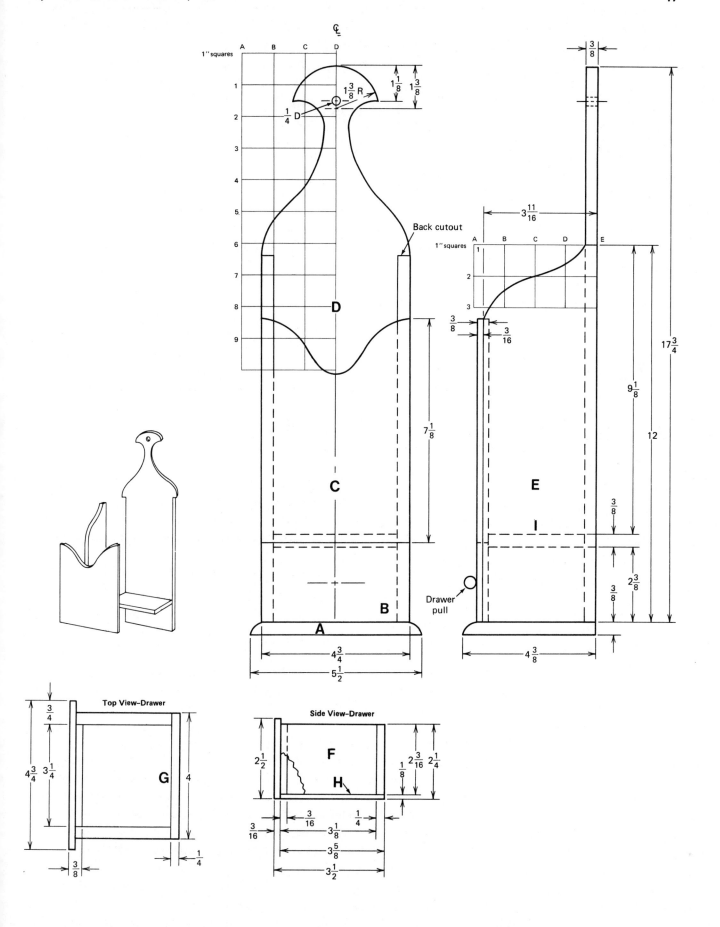

Back cutout

Drawer pull

Top View–Drawer

Side View–Drawer

PROJECT
3

Shaker-Style Bench

Several Shaker-style pieces appear in this book. These highly functional designs are usually simple to construct. In addition, they fit in with almost any decor, but of course they are ideally suited for use with Early American pieces.

The largest part of this project is the bench top. It is 1 in. thick, 10 in. wide, and 36 in. long. In true Shaker furniture this piece would be fashioned from a single wide pine board. Although a single board is an option here, the use of three narrower boards, with heartwood alternated, joined edge to edge with dowels, will help to keep the top flat.

The legs can be cut from a single piece of $\frac{3}{4}$-in. by 10-in.-wide lumber 33 in. long. This will yield two $16\frac{1}{2}$-in.-long legs. Another option is to combine three narrower pieces edge to edge, as was done with the top. The two legs can then be fastened together face to face and handled as a single piece.

The circular cutout can be made with a saber saw or a bandsaw. The two $\frac{3}{4}$-in.-wide cutouts for the horizontal support can be cut with hand tools after a row of $\frac{3}{4}$-in.-diameter holes has been bored along the centerline of the slot. If available, a bandsaw can be used to cut down to a single hole bored at the bottom of the slot. Then the slot can be squared up with a file.

The horizontal supports can be cut from a single piece of lumber $\frac{3}{4}$ in. by $3\frac{1}{2}$ in. wide and 20 in. long. After cutting the piece in half, the two segments are fastened together face to face and treated as a single piece.

Assembly is simple. Insert the horizontal supports into the slots cut into the ends of the legs and bore two $\frac{3}{8}$-in.-diameter holes into the edges of the legs and into and through the horizontal supports. Insert glue-covered dowels into the holes and trim flush.

The ends of the legs fit into dadoes cut into the bottom face of the bench top. Screws, $1\frac{1}{4}$-in. No. 12s, are driven through the underside of the horizontal supports and into the bottom face of the bench top. Glue can be used in addition to the screws if desired. However, without glue the bench can be quickly knocked down by unscrewing the screws.

Shaker-Style Bench

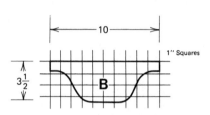

Letter	Description	Number of pieces	Thick-ness	Width	Length
A	Leg	2	3/4	10	16–1/2
B	Horizontal support	4	3/4	3–1/2	10
C	Top	1	1	10	36

Notes
1. Material: pine or cherry.
2. Top, C, made up of three pieces of equal–width edge joined with dowels (3/8–in. diameter) 2–in. from ends at 12–in. intervals, heartwood reversed on alternate piece 1.
3. Top, C, fastened to part B with screws 1–1/4–in. No. 12 flathead screws.
4. Leg, A fits into 3/4–in. x 1/2–in. deep dado in top of part C.
5. All outside corners rounded to 1/4–in. radius.

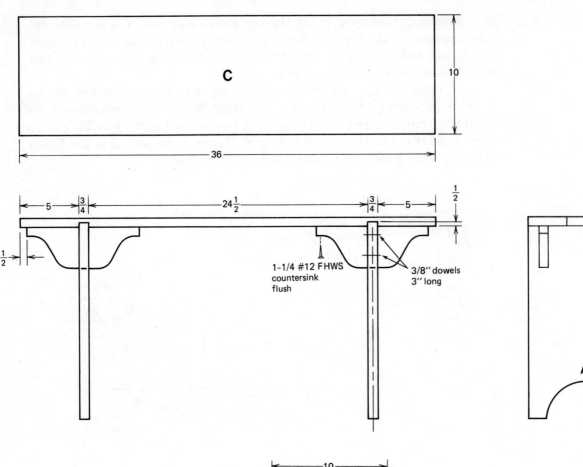

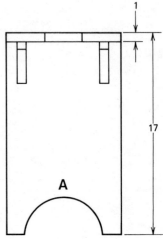

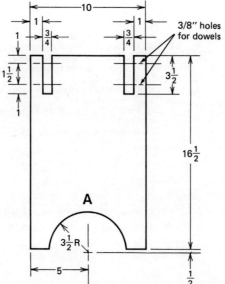

C

10

36

5   3/4   24 1/2   3/4   5   1/2

1/2

1-1/4 #12 FHWS
countersink
flush

3/8″ dowels
3″ long

1

17

A

10

1   1   1

1   3/4   3/4

1 1/2

3 1/2

3/8″ holes
for dowels

16 1/2

A

3 1/2 R

5

1/2

PROJECT
4

Kitchen Wall Rack

This useful project can serve the dual roles of wall rack, for hanging up pots, and storage shelf. The overall dimensions given in the drawing will look good and provide a well-proportioned piece. If desired, the dimensions, especially the width, can be changed to suit individual space and storage requirements.

The sides of the piece are shown in the drawing as being fabricated from three pieces joined edge to edge. Another possibility is to use hardwood-veneer-covered plywood for the sides, with the edges banded with a strip of wood that matches the face veneer of the plywood.

The dowels are shown coming through the sides; however, the dowels can be cut flush with the sides or they can be fitted into blind (not through-bored) holes bored into the sides with a flat-bottomed bit such as a Forstner bit.

The sides are made up of three pieces, $\frac{3}{4}$ in. by $2\frac{1}{2}$ in. by 28 in., and two $\frac{3}{4}$-in. by $4\frac{1}{4}$-in.-wide pieces, one about 4 in. long and one about 12 in. long. Of course, since two sides are required, two pieces of each size must be cut. The pieces are doweled together edge to edge. After squaring the long straight edge on each side assembly, they can be fastened together face to face and treated as a single piece. The fastening can be done with double-sided carpet tape or with finishing nails driven into areas where they will not show up later as nail holes. The next step is to cut the ends of the side assemblies square to the straight edge and at the same time to reduce the side assembly to its finished length, 28 in. The sides are now cut to and planed to their final width, 11 in.

After the curved section is laid out, the curve can be cut on the bandsaw or with a saber saw.

Hole centers are located and holes are bored into one face of the two assembled sides until the bit's center point emerges through the opposite face. Then the piece is flipped over and the bit inserted into the hole made when the bit broke through, and boring is completed. This will help to reduce splitting. Of course, a depth stop would be used on the bit and all holes would be bored into the first face before reversing the piece to complete the boring operation.

The top is made up of three pieces $3/4$ in. by 4 in. by 34 in. They are fastened together edge to edge with dowels, and with the heartwood reversed on each piece. A single piece can be substituted, but cup warping may become a problem.

Two horizontal rails measuring $3/4$ in. by $2\frac{1}{2}$ in. by 32 in. must be made. These are fastened to the sides, as shown in the drawing, using two $3/8$-in. dowels inserted into holes bored through the sides and into the ends of the horizontal rails.

The top can be attached to the upper ends of the sides with dowels or with flathead wood screws which are counterbored and covered with plugs (short lengths of dowel).

In the actual assembly it would be a good idea to insert all of the $5/8$-in. by 34-in. dowels into both sides before attaching the other pieces. The dowels will go in more easily before the sides are attached to the other project parts. If desired, all the outside corners can be rounded over with a router and rounding-over bit or with a block plane and a spokeshave or file.

The shelf should be attached to the wall by screwing through the upper horizontal rail and into the wall studs behind the wall rack.

Kitchen Wall Rack

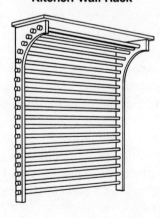

Letter	Description	Number of pieces	Thick- ness	Width	Length
A	Sides	2	3/4	11	28
B	Top	1	3/4	12	34
C	Horizontal rails	2	3/4	2½	32
D	Dowels	20	5/8	5/8	34

Notes
1. Material: oak; birch dowels.
2. Sides made up of 3 pieces: one piece 3/4-in. x 2-1/2-in. x 28 in., one piece 3/4-in. x 4-1/4-in. x 4 in. joined edge to edge with 3/8-in. dowels.
3. Horizontal rails, C, fastened to sides, A, with 2-3/8-in. dowels centered on C and 1/2-in. in from edges of part C.
4. Outside edges of parts A and B may be shaped if desired.

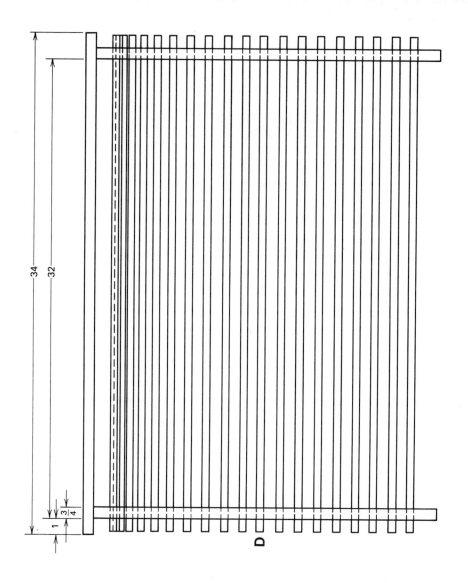

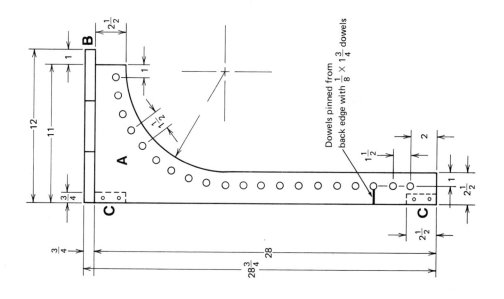

Dowels pinned from $\frac{1}{8} \times 1\frac{3}{4}$ dowels

PROJECT
5

Shaker-Style Step Stool

This unique piece would be at home in any Early American setting. The stool has a long handle which steadies the user as he or she steps up on the stool and which was used to hang the stool on a wall peg when not in use. This was the practice of the Shakers, who installed rows of wooden pegs on the wall to support chairs and other furniture when they were not needed.

The largest piece of the project is the back assembly, which is made up of three pieces. The back center is fashioned from a single piece measuring $\frac{3}{4}$ in. by $3\frac{3}{4}$ in. by 34 in., and two smaller pieces, the back right and left, each measuring $\frac{3}{4}$ in. by 9 in. These are fastened together edge to edge with $\frac{3}{8}$-in.-diameter dowels. The heartwood in each piece is reversed to reduce warping. An alternative method of construction is to use one piece to make the back.

Before assembling the three back pieces, the long double taper should be cut on the center piece. This can be done on a table saw with a taper jig, or the piece can be ripped oversized and planed to final width.

The step is also fabricated from three pieces. The center piece is $\frac{3}{4}$ in. by $3\frac{3}{4}$ in. by 13 in. and the two side pieces are $\frac{3}{4}$ in. by $1\frac{7}{8}$ in. by 13 in. These pieces are joined edge to edge with dowels in much the same way that the back was. Before gluing the three pieces together, cut out the $\frac{3}{4}$-in.-wide, $3\frac{3}{4}$-in.-long opening in the center section of the step. This will receive the back handle later. Also cut the dadoes into the bottom faces of the smaller pieces on each side of the center step piece. Then reassemble all the pieces.

The front leg of the stool is made up of three pieces (parts A, B, and C) in a similar way. This leg is fastened to the step with a multiple dovetail joint, as indicated in the drawing.

Notice that added strength is provided by a ³⁄₄-in.-diameter dowel rail, J. This dowel fits into holes bored through the front and rear legs of the bench. The ends of the dowel are cut with a backsaw so that a groove about ³⁄₄ in. deep is cut in the *same direction* as the grain of the dowel. After the project is assembled, hardwood wedges are coated with glue and driven into the ends of the dowel. This causes the dowel to expand, binding it to the leg. When the glue sets, the excess portion of the wedge is removed. This type of joint is known as a foxed joint.

Shaker-Style Step Stool

Letter	Description	Number of pieces	Thickness	Width	Length
A	Front left	1	3/4	1-7/8	9
B	Front center	1	3/4	3-3/4	9
C	Front right	1	3/4	1-7/8	9
D	Back left	1	3/4	1-7/8	9
E	Back center	1	3/4	3-3/4	34
F	Back right	1	3/4	1-7/8	9
G	Step left	1	3/4	1-7/8	13
H	Step center	1	3/4	3-3/4	13
I	Step right	1	3/4	1-7/8	13
J	Rail 3/4-in. dowel	1	3/4 D	3/4	11

Notes
1. Parts D, E, F—A, B, C—G, H, I—doweled edge to edge, 3/8-in. dowels 1-1/2-in. in from the ends of the shaped pieces.
2. Ends and edges of the step are shaped using a rounding-over router bit.
3. Parts D and F fit into 3/8-in.-deep dado cut into the lower face of parts G and I.
4. Rail dowel J, ends cut through center 1/4-in.-wide 3/4-in.-deep wedges driven in at assembly.**

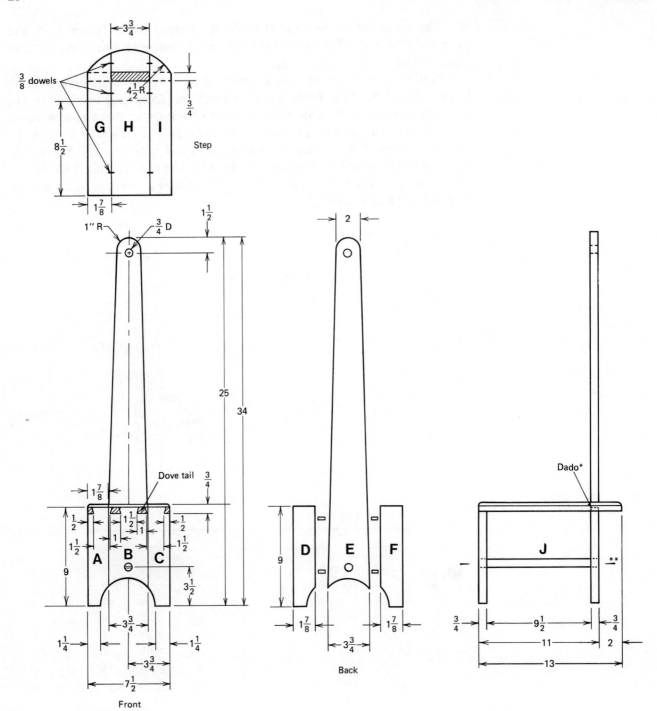

$\frac{3}{8}$ dowels

$3\frac{3}{4}$

$4\frac{1}{2}$R

$\frac{3}{4}$

G H I

$8\frac{1}{2}$

$1\frac{7}{8}$

Step

1" R

$\frac{3}{4}$ D

$1\frac{1}{2}$

2

25

34

$1\frac{7}{8}$

Dove tail

$\frac{3}{4}$

$\frac{1}{2}$

$1\frac{1}{2}$

1

1

$\frac{1}{2}$

$1\frac{1}{2}$

$1\frac{1}{2}$

A   B   C

9

$3\frac{1}{2}$

$1\frac{1}{4}$

$3\frac{3}{4}$

$1\frac{1}{4}$

$3\frac{3}{4}$

$7\frac{1}{2}$

Front

9

D   E   F

$1\frac{7}{8}$

$1\frac{7}{8}$

$3\frac{3}{4}$

Back

Dado*

J

$\frac{3}{4}$

$9\frac{1}{2}$

$\frac{3}{4}$

11

2

**

13

PROJECT
6

Bandsawed Jewelry Box

This project was designed by Jack N. Ottomanelli, who made the original out of zebrawood. Any suitable wood can be used.

The first step in making this piece is to draw a full-sized pattern of the front view of the project. Then cut out 8 pieces measuring ¾ in. by 12 in. wide by 10 in. long and one ¼-in.-thick piece 12 in. wide and 10 in. long, which will become the front of the assembly. In the drawing these pieces are identified by the letters A, M, F, G, H, I, J, K, and L. Part A is the ¼-in.-thick piece and part L is the back piece.

These rectangular pieces are stacked up face to face with part L on the bottom of the stack and part A on the top of the stack. All are glued together face to face. *The front and back, A and L, have a piece of brown kraft paper (heavy wrapping paper) in their glue joints.* This will enable the front, A, and the back, L, to be split away from the assembly later.

After the glued-up assembly has cured overnight, the clamps are removed and the full-sized pattern is used to draw the project shape on the face (part A) of the assembly. Now the *outline* of the shape is cut out using the bandsaw. The outside edges of the assembly are smoothed up using files and sanding equipment.

At this point the back piece, L, is split away from the assembly using a knife and chisel which is inserted into the glue joint between parts L and K, splitting the layer of kraft paper which was inserted when the assembly was glued up.

Now the internal cuts are made into the remaining glued-up pieces. Notice that the cut should begin and end in the center of the base section of the assembly. When this cutting is complete, parts B, E, D, and C will fall free of the assembly. The ¼-in. layer of each of these pieces is now split away from each large cutout piece and serves as a drawer front.

The detailed drawings of the drawers indicate the construction of each. Drawer C is made by taking the cutout section that remains after the ¼-in. front is split away and boring two 1-in.-diameter holes ¾ in. deep for rings. A shallow curved section is bandsawed out to form a recess for other jewelry. When these operations are finished, the front is reattached and flocking or felt is applied to the upper surfaces of the drawer. Note that a ¼-in.-thick back is attached. Each of the other drawers, B, D, and E, are made in a similar way.

When all the drawers are complete, the back piece, L, is glued back onto the assembly. Wooden drawer pulls are attached and the outside of the piece has several coats of polyurethane varnish applied to provide a glossy finish.

The result is an eye-catching "fun" jewelry box.

Bandsawed Jewelry Box

Letter	Description	Number of pieces	Thick-ness	Width	Length
A	Front of cabinet	1	1/4	12	10
B	Drawer				
C	Drawer				
D	Drawer				
E	Drawer				
F	One of nine pieces making up cabinet	1	3/4	12	10
G	One of nine pieces making up cabinet	1	3/4	12	10
H	One of nine pieces making up cabinet	1	3/4	12	10
I	One of nine pieces making up cabinet	1	3/4	12	10
J	One of nine pieces making up cabinet	1	3/4	12	10
K	One of nine pieces	1	3/4	12	10
L	Back of cabinet	1	3/4	12	10
M	One of nine pieces making up cabinet	1	1/2	12	10
O	Piece used for individual drawer backs	1	1/4	12	10

Notes
1. Material: zebrawood or maker's choice.
2. See separate detail drawings of the drawers.*
3. Pieces A M F G H I J K L, cut to size and glued–up face to face to form one large piece. Kraft paper is used between parts A and M and between parts K and L when the block is glued up.
4. After glue sets, block is squared to rough dimensions–6 in. x 12 in. x 10 in.
5. Outline shape of block is cut out using a bandsaw.
6. Cabinet back, L, is removed by carefully separating part L from K by splitting Kraft paper between the pieces with a stiff–bladed knife and chisel. Work from block bottom.
7. Drawer-on front shapes are cut out through the remaining block made up of parts A, M, F, G, H, I, J, and K.
8. The cutout shapes B, C, D, and E are used to make the drawers B, C, D, and E. See individual drawings. The front of each of these blocks is removed as per note 6. The larger pieces are used to make the inside part of each drawer. The smaller parts are used for the drawer fronts.
9. "L" is glued back onto "K" of the original block at this point.

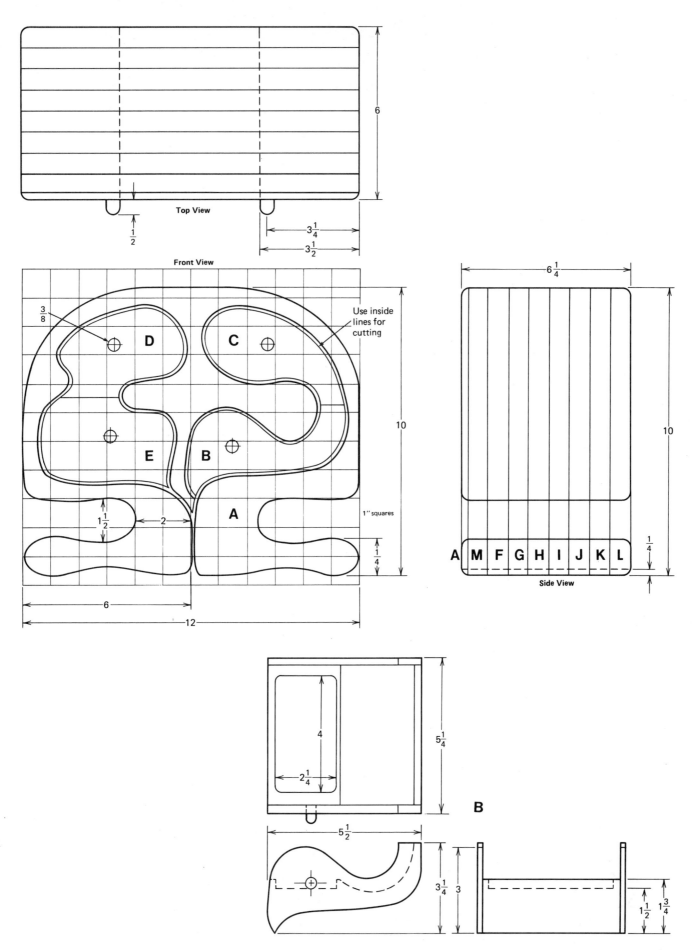

Top View

Front View

$\frac{3}{8}$

D   C

Use inside
lines for
cutting

E   B

10

$1\frac{1}{2}$ 2

A

$\frac{1}{4}$

1" squares

6

12

$\frac{1}{2}$

$3\frac{1}{4}$

$3\frac{1}{2}$

6

$6\frac{1}{4}$

10

A M F G H I J K L

$\frac{1}{4}$

Side View

4

$2\frac{1}{4}$

$5\frac{1}{4}$

B

$5\frac{1}{2}$

$3\frac{1}{4}$ 3

$1\frac{1}{2}$ $1\frac{3}{4}$

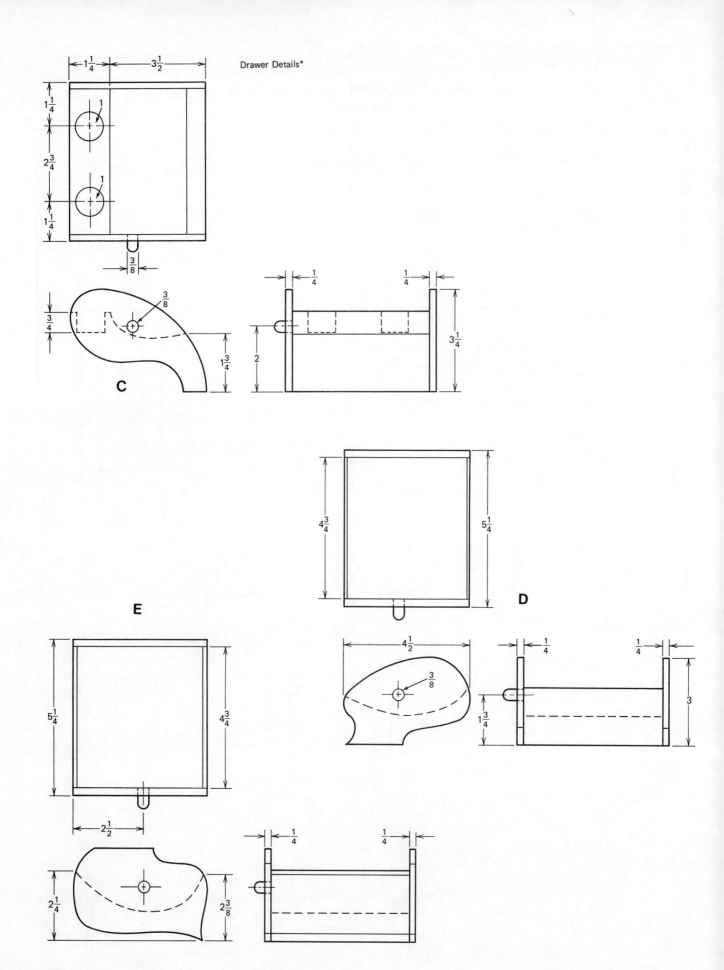

Drawer Details*

C

D

E

PROJECT
7

Pie Table

This small table is well suited as a lamp or plant table. Traditionally, this type of table is often fitted with three legs instead of four. The three-legged version requires a round or hexagonally shaped post bottom. In addition, the joinery required, usually a spline dovetail, is difficult to make with the equipment available in the typical nonproduction shop.

The four-legged version is easier to fashion because the post is made square in cross section during the squaring-up process. In addition, legs can be clamped to the post, one at a time, because the opposite side of the post is flat and parallel to the edge of the leg being clamped.

The tabletop is cut from a square approximately $14\frac{1}{2}$ in. by $14\frac{1}{2}$ in. The two inside pieces measure $3\frac{1}{2}$ in. in width, while the two outside pieces measure $3\frac{3}{4}$ in. in width. When the 14-in. circle is cut from this oversized square, all four of the top pieces will measure $3\frac{1}{2}$ in. in width. The top pieces are doweled together edge to edge with the heartwood on each piece reversed. The circular top can be cut out on a bandsaw using a circle cutting jig. This consists of a board clamped to the saw table with a nail driven through the center of the uncut tabletop into the clamped-on board, in line with the saw blade and 7 in. from it. An alternative method is to use a power router fitted with a trammel attachment and a small-diameter straight cutting bit. The circle is cut by making several passes around the piece, lowering the bit a little more each time. A 4-in.-wide piece of wood about 10 in. long and equal in width to the router base can be screwed to the router base in place of the plastic disk usually attached to it. This board becomes a trammel when a nail is driven through it 7 in. from the bit. The edge of the tabletop is shaped using a piloted router bit of suitable shape.

The shape shown for the table legs provides a flat area that will be parallel to the flat surface on the post when the legs are attached. This will make clamping of the legs to the post easier.

Since the legs are identical in shape, they can be fastened together after they are laid out on the mitered pieces shown in the drawing, and cut and sanded as a unit.

The table legs can be attached to the post with mortise-and-tenon, blind, or through dowel joints. If dowels are used, the holes for the dowels are made in the legs first. Then one matching hole is made in the post using a dowel center inserted into one of the holes bored in the leg to locate its center. After the first hole is bored in the post, a dowel center is put into the second hole in the leg, while a short dowel is put into the other leg hole and the first hole bored in the post. This dowel helps to stabilize the leg as it is positioned. Then the leg is pressed into the post and the dowel center locates the second hole in the post.

If a mortise-and-tenon joint is used, the legs should not be cut to shape until after the tenon is cut on the end of the leg. To do otherwise would be dangerous.

The tabletop is supported by a "cross-shaped" half-lapped assembly, G and F. The assembly has a hole bored through its center. This hole is equal in diameter to the cylinder turned on the top of the post. To ensure a good fit, the cylinder should be cut slightly oversized. It is reduced to final size by forcing the cylinder into a hole bored into a piece of hardwood by the same bit that was used to make the hole through the cross-shaped table support.

Pie Table

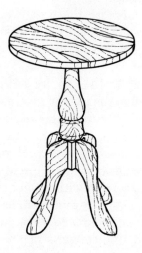

Letter	Description	Number of pieces	Thick-ness	Width	Length
A,B,C,D	Legs	4	3/4	6	9
E	Top	4	3/4	3-1/2	14
F,G	Top support	2	3/4	4	12
H	Post	1	2-1/4	2-1/4	14-1/4

Notes

1. Material: maker's choice.
2. Solid wood top made up of four pieces joined edge to edge with dowels.
3. Dowels: top and legs, 3/8 in. in diameter.
4. Top edge may be shaped using router.
5. Leg shape allows for clamping legs to post while glue cures.
6. Screw holes in table support, (G), going across the grain must be oval to allow for top movement across grain.
7. 1-1/4 in. No. 10 flathead wood screws—8. Hold top to support. Do not glue top to supports; everything else glued.

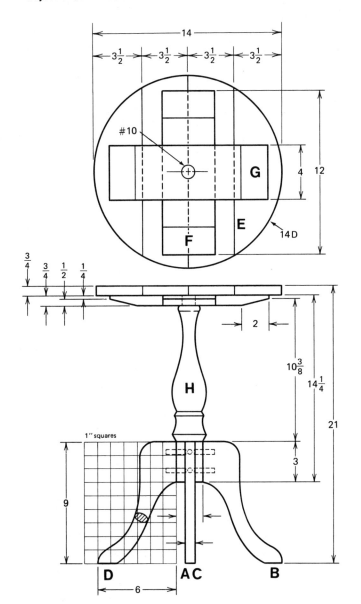

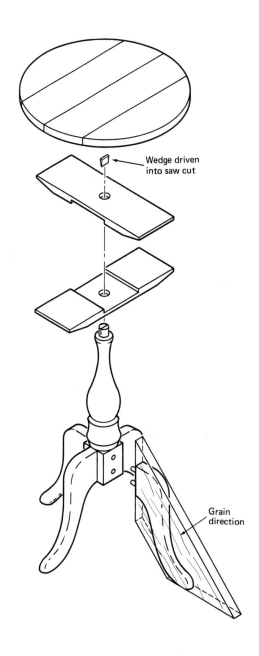

Wedge driven
into saw cut

Grain
direction

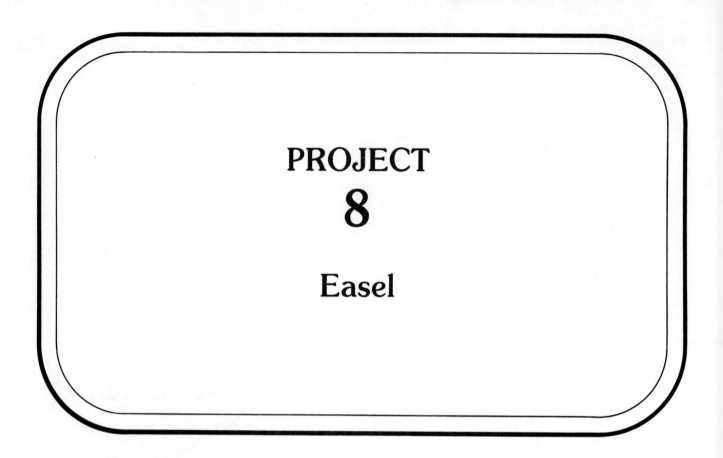

PROJECT
8

Easel

This easel is a simple, yet functional version of the classic studio fixture. Although somewhat shorter than the typical studio model, it will function efficiently as a holding or display device for all but the largest piece of art.

An easel is a variation on the tripod. In this design the base, A, provides two of the three supports of the tripod. The rear vertical support, D, makes up the third leg of the tripod. Parts A and D are essentially rectangles. The front vertical piece, B, is almost identical to part D, and both pieces can be cut from a single piece 5 in. wide (trimmed down from a 1 × 6). Part B has a slot measuring 40¼ in. long by 1¼ in. wide. This slot can be ripped by making two cuts that intersect a 1¼-in.-diameter hole bored at the end of the slot. The slot is capped by part C, which can be cut from a rectangular piece 5 in. by 2¾ in. long. Part C can be fastened to the top of the slotted front vertical piece with 1½-in. No. 8 flathead screws or by using two ⅜-in.-diameter dowels.

The bottom horizontal canvas holder, G and H, and the top horizontal canvas holder, L and M, are very similar in construction. Each consists of a tray and back, recessed on the table saw to provide a wide slot into which the canvas fits. The slot should be left rough-cut, as this roughness will help to stabilize the canvas. The back is glued to the tray and forms an L-shaped piece which holds the canvas.

The horizontal canvas holders, GH and LM, can slide up and down in the slot in the front vertical support, B. A projection on the back of parts M and H, which is $3/8$ in. thick, mates with a similar projection on part I, which is bolted to it with a machine screw and wing nut. The projection on part I is only $1/4$ in. thick. Thus, when the two projections in the slot are mated, they squeeze against the front vertical support and stay in place. Part I, two of which are required, can be cut from a single piece of $3/4$-in.-thick material.

The front and rear vertical supports are connected at the upper end of the easel by means of a butt hinge. These pieces are held at a fixed distance apart by a movable horizontal brace formed by the movable horizontal brace side, E, and the dowel connector, J. The front dowel connector fits into a $3/8$-in.-diameter hole bored from edge to edge on part B 1 in. below the slot. The rear dowel connector, J, fits into a grooved block, F, which is screwed onto the rear vertical support, D. This configuration allows the horizontal brace to fold up flat between parts B and D when the easel is folded up. Two glides should be fastened to the lower edge of the base, A, to provide two-point contact.

Easel

Letter	Description	Number of pieces	Thick- ness	Width	Length
A	Base	1	3/4	3	24
B	Front vertical piece	1	3/4	5	60-1/4
C	Top connector	1	3/4	5	2-3/4
D	Rear vertical support	1	3/4	5	61-1/2
E	Movable horizontal brace side	2	1/2	1-1/4	22
F	Rear brace block	1	1-1/8	1-1/2	5
G	Bottom horizontal canvas holder	1	3/4	2-1/2	24
H	Back of bottom horizontal canvas holder	1	1/2	2-1/4	24
I	Rear clamping piece of canvas holders	2	1-1/4	2-1/4	5
J	Dowel connector of horizontal brace	2	3/8 D	3/8 D	6-1/2
K	Carriage bolt and washers	2	1/4	1/4	3
L	Top horizontal canvas back holder	1	3/4	2-1/2	9-1/2
M	Back of top horizontal canvas holder	1	1/2	2-1/4	9-1/2

Notes
1. Material: poplar.
2. F–G and L–M glued and screwed with 3 1-in. No. 8 flathead wood screws.
3. Edges and ends parts A and D shaded with beading or corner-rounding router bit.

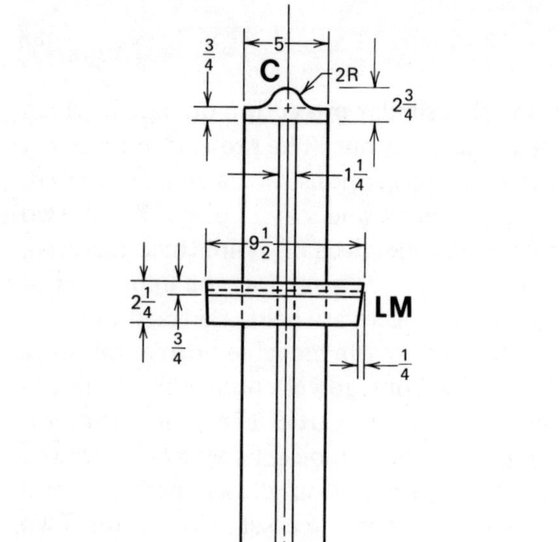

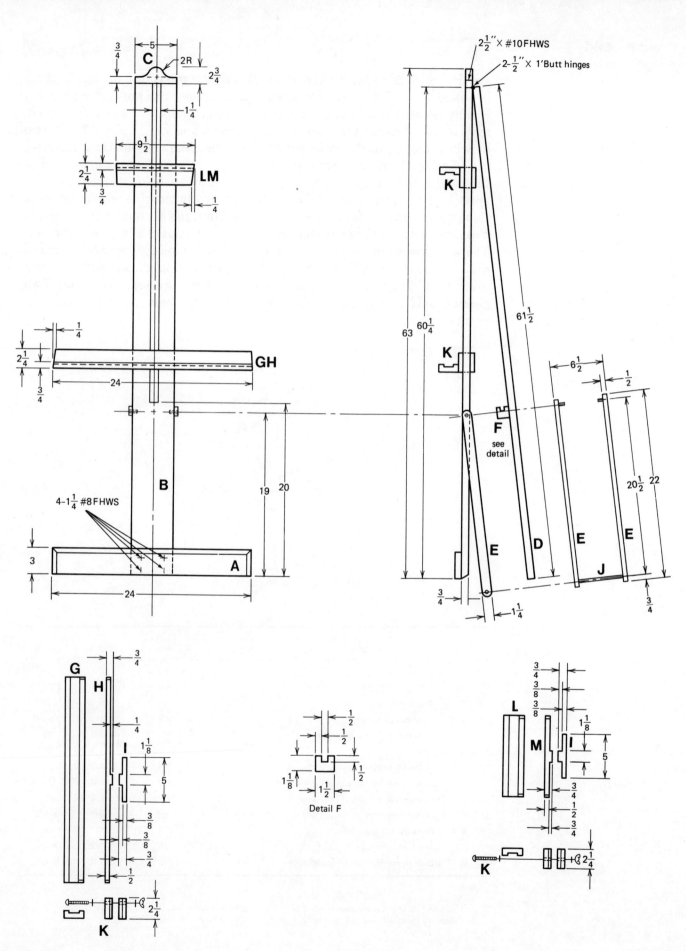

5
3/4
C
2R
2 3/4
1 1/4
9 1/2
2 1/4
LM
3/4
1/4
1/4
2 1/4
GH
3/4
24
B
19
20
4-1 1/4 #8 FHWS
3
A
24

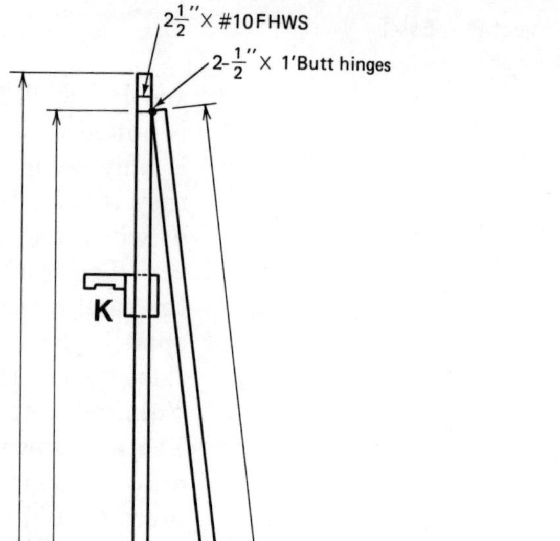

2 1/2" × #10 FHWS
2-1/2" × 1' Butt hinges
K
K
63
60 1/4
61 1/2
F
see detail
6 1/2
1/2
20 1/2
22
E
D
E
E
J
3/4
3/4
1 1/4

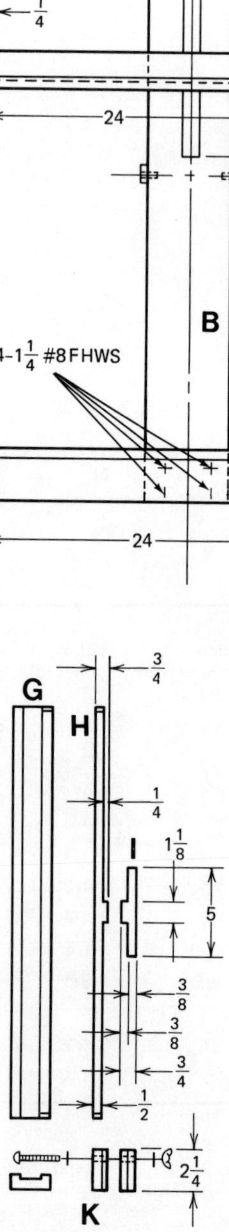

3/4
G
H
1/4
1 1/8
I
5
3/8
3/8
3/4
1/2
2 1/4
K

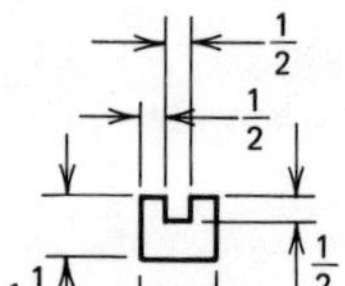

1/2
1/2
1/2
1 1/8
1 1/2

Detail F

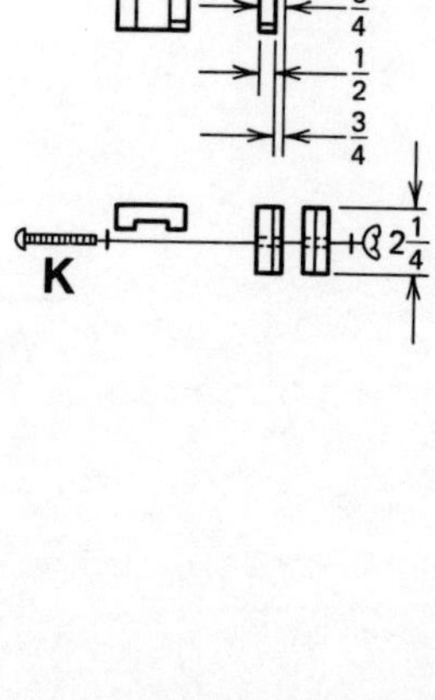

3/4
3/8
3/8
L
1 1/8
M
I
5
3/4
1/2
3/4
K
2 1/4

PROJECT
9

Coffee Table

This coffee table is of contemporary design. It consists of a 28-in. by 58-in. top divided into three sections. The center section is 16 in. wide and is made of slate or wood covered with a plastic laminate like Formica® brand.* The two sections on each side of the slate are made of the wood used to make the table: teak or walnut. This arrangement not only adds visual interest but provides a mar-resistant surface which is highly functional on a coffee table.

The tabletop is supported on two inverted-U-shaped components which are joined together by top and bottom rails. The two top rails support the table and provide additional stability for the U-shaped leg assemblies.

The tabletop's outer pieces, E, are rabbeted (see the cross-sectional view with the end cap removed) to receive the top center piece, G, which is a piece of ¾-in.-thick plywood. These parts are screwed and glued together. The screws are 1-in. No. 8 flathead wood screws, and they are driven through the plywood and into the rabbeted side pieces. The ends of the table are capped with a 1-in.-thick end-cap piece, F. The edges and ends of the table-top are shaped into a half-round with a power router or plane after the end caps are in place.

The U-shaped leg assemblies consist of two vertical components: legs, C, and a horizontal component, the bottom leg support, B. These parts are

*Formica® brand and Colorcore® brand plastic laminates are registered trademarks of the Formica Corporation.

joined using a splined miter joint. The pieces are cut at a 45° angle on a table saw or with a radial arm saw, if available. Then a ¼-in.-wide groove, ½ in. deep, is cut into the end of each piece on center. This forms a slot into which a spline ¼ in. thick, 6 in. wide, and 1 in. long, fits. (See the exploded view on the drawing.) The spline strengthens the joint and locates the pieces relative to each other during assembly.

The legs are fastened to the top rails with plugged screws, driven into anchor dowels inserted into the rail edges. (See the detail on the drawing.) The leg support is attached to the bottom rail using two blind dowels as shown in the drawing.

If teak is used for this piece, care must be taken in gluing. Wash all surfaces to be glued with lacquer thinner, in a well-ventilated area, to remove excess resin. When the piece is dry, mix up *fresh* casein glue and assemble. It is a good idea to test the glue on scrap teak before gluing up the project.

Teak is somewhat abrasive and it tends to dull cutting edges quickly unless they are carbide tipped.

Coffee Table

Letter	Description	Number of pieces	Thick-ness	Width	Length
A	Bottom rail	1	1–1/4	3–5/8	42
B	Bottom-leg support	2	1–1/4	3–3/4	24
C	Leg	4	1–1/4	3–3/4	14–3/4
D	Top rail	1	1–1/4	2	42
E	Top outer piece	2	1–1/4	6	56
F	End cap	2	1	1–1/4	28
G	Top center piece (plywood)	1	3/4	21	56
H	Slate formica (black) may be substituted	7	1/2	8	16
I	Spline (see leg bottom joint, detail C–B)	4	1/4	6	1

Notes
1. Material: teak or walnut.
2. If Formica is used part G will have to be thickened by 7/16 in. in the area under the Formica.
3. Part F doweled to part E with 2-1/2-in. dowels into each part E.
4. Oil finish suggested.
5. Casein glue after washing surfaces to be glued with lacquer thinner.

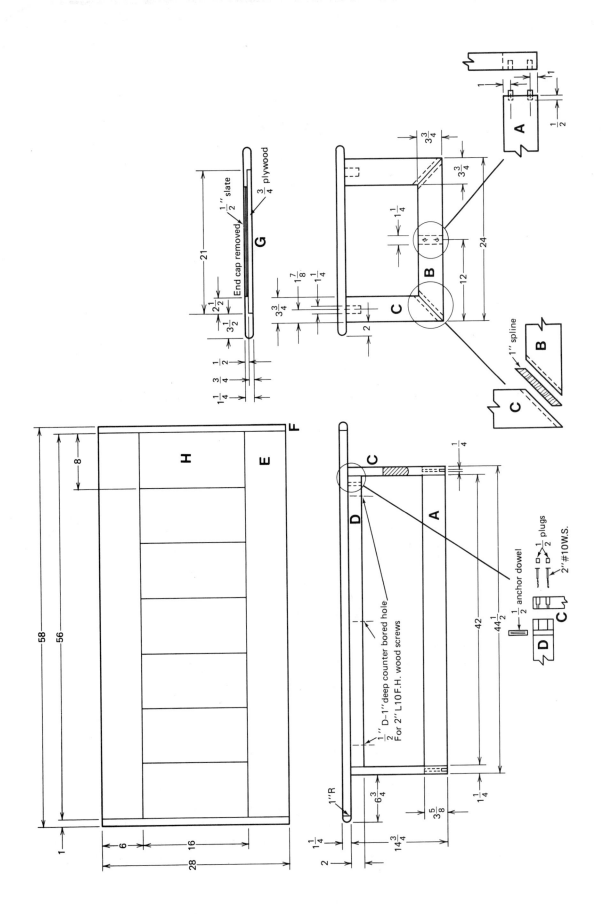

PROJECT
10

Toolbox

This toolbox, derived from an original design by Beverly J. Bloom, is designed to hold the basic hand tools of a carpenter. It is large enough to hold two handsaws in the upper section, G. The saw rack, formed by parts F and G, can be lifted out or slid forward or back to provide access to the sliding storage box, H, or to the storage area under the storage box. With the storage box slid to one side, space is available in the area under and next to the box for some models of portable circular saw.

The swing-out section has covered shelves which can be used to hold planes, folding rules, chisels, and similar equipment. The shelf in this section can be eliminated or relocated if desired.

The box should be made out of cabinet-grade hardwood plywood. Most of the parts are joined together with rabbet or rabbet dado joints. These are glued for maximum strength. The end piece, A, is cut from a piece of ½-in.-thick plywood measuring 11 in. by 18 in. The swing-out section end, O, is cut from part A after it has been squared to final dimensions.

The top, D, and swing-out section top, L, are cut from a single piece measuring ½ in. by 11 in. wide by 32½ in. long.

The swing-out section back, Q, and toolbox front, P, are cut from a single piece of ½-in.-thick plywood 18 in. by 32 in. long.

The back, C, and the front, P, are fastened to the box bottom, B, and the box top, D, with glue and ¼-in. through dowels at 4-in. intervals. These dowels can be blind, if desired.

After the project parts are cut to final size and all joints are completed, assemble the bottom, B, the top, D, and the ends, A and E.

Assemble the swing-out section from swing-out section back, Q, shelves, M, top, L, cover, N, and ends, O and K. The cover, part N, is attached to the top shelf, M, with a piano hinge. A magnetic catch should be installed to hold the cover, N, in place against the swing-out section top, L.

A piano hinge attaches the swing-out section to the toolbox. The hinge is screwed to the toolbox front, P, and the swing-out section back, Q. This hinge is installed after both major components, the toolbox and the swing-out section, have been assembled.

The saw rack runners, I, are glued and screwed into place after the toolbox is assembled. Storage box runners, J, are also installed after assembly of the major components.

The sliding storage box is shown constructed with glued butt joints. Finger joints or rabbet dadoes may be substituted for side and end joints. The bottom can be fitted into grooves cut into the sides and ends of the box for added strength.

The toolbox should be mounted on four rubber-tired swivel casters of appropriate size. Handles should be installed on the sides of the box as shown in the exploded isometric view of part A. Luggage-type locking hasps can be installed at the top of the toolbox to keep it securely closed.

Another option allows replacement of the handsaw rack with drawers for other tools.

The box should be finished inside and out with spar varnish for protection against the weather.

Toolbox

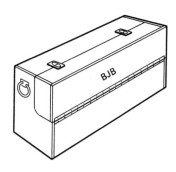

Letter	Description	Number of pieces	Thick-ness	Width	Length
A	Left end	1	1/2	11	18
B	Bottom	1	1/2	10-1/2	32-1/2
C	Back	1	1/2	17	32
D	Top	1	1/2	5-1/2	32-1/2
E	Right end	1	1/2	11	18
F	Saw rack bottom	1	3/4	4-1/2	32 *
G	Saw rack support storage box	2	3/4	4-1/2	5-1/2
H	Sliding storage box 1/4-in. bottom, 1/2-in. sides	1	3	9-12	16-1/2
I	Saw rack, runner	2	1/2	1/2	10-1/2
J	Storage box runner	2	1/2	1/2	32
K	Swing-out section right end	1	1/2	5-1/2	9
L	Swing-out section top	1	1/2	5-1/4	32-1/2
M	Swing-out section shelf (middle) (top)	2	1/2	4-1/2	32-1/2
N	Swing-out section cover	1	1/2	6	32
O	Swing-out section left end	1	1/2	5-1/2	9
P	Toolbox front (lower half)	1	1/2	9	32
Q	Swing-out section back	1	1/2	9	32

Notes
1. Material: hardwood cabinet plywood recommended.
2. Additional requirements: 2 32-in. piano hinges; 2 4-in. carrying handles. Optional; 4 casters.
3. Ends A and O, should be made from a single piece 1/2-in. x 11 in. x 18 in. then cut into two pieces.
4. Back, C, and P, front, glued to box last.
* 5. This piece F will be trimmed to allow a loose fit.

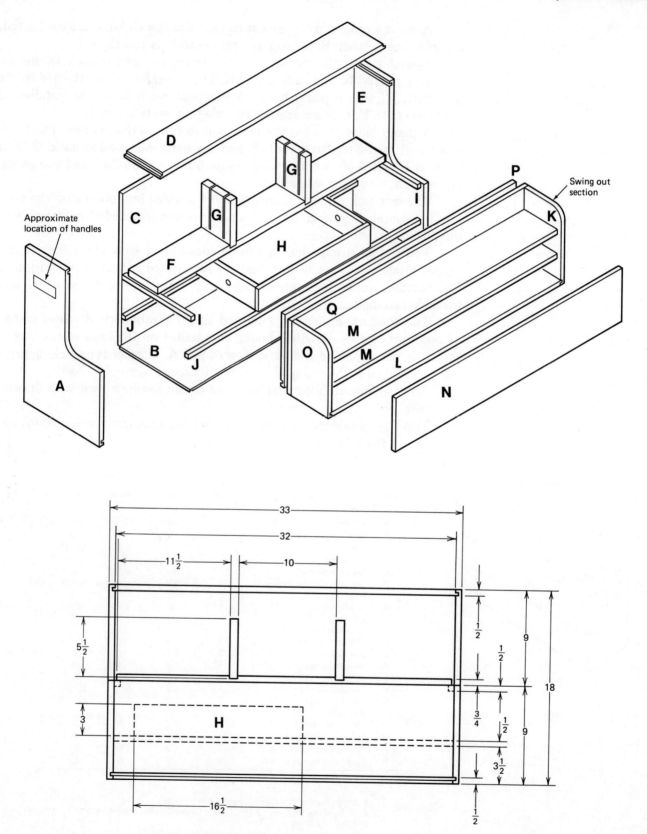

Approximate location of handles

Swing out section

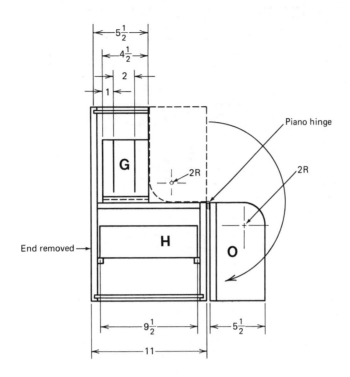

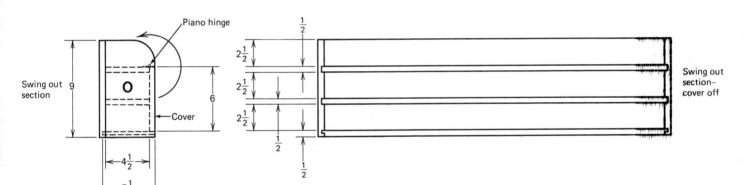

PROJECT
11

Dining Table

This project consists of a 42-in.-diameter round tabletop supported by a four-posted column and a cross-shaped base. The table is 29 in. high. It can be used in the dining room or kitchen, and it will seat four with great comfort and six with slightly less comfort.

The top is made up of seven pieces joined edge to edge with dowels. Each piece measures 1½ in. by 6 in. by 42 in. An alternative construction technique would utilize two pieces of ¾-in.-thick plywood, the top piece veneered with teak or walnut and edged with teak or walnut.

A top of this size can be cut to a diameter of approximately 42½ in. using a power saber saw. The final 42-in. dimension can be cut using a power router mounted on a trammel and fitted with a carbide-tipped straight bit. The rounding over can be done with a rounding-over bit and a power router after the final diameter is reached.

The base is fabricated by cross-lapping two pieces, D. Parts D can be shaped by utilizing the split-turning method on a wood lathe. Two pieces measuring 2 in. by 4 in. by 32 in. are glued together with a piece of kraft paper in the glue joint. For added stability, two temporary wood screws can be driven into the straight center section of the piece. The piece is mounted between centers on the lathe and the rounded tapered section of the piece is shaped. This is 12 in. in length as measured from the ends of the piece. After sanding, the piece is removed from the lathe. The screws are removed and the two pieces split apart along the glue line using a chisel. The kraft paper will actually split. If a lathe is not available, these pieces will have to be shaped with a plane and files.

The table supports, B, are cross-lapped as shown in the drawing. In addition, they are dadoed to receive the upper ends of the posts, C.

The posts, C, are bolted to the base and table supports with lag bolts as shown in the drawing. Note that the lag bolts are screwed into anchor dowels located in the bottom and top edges of the posts. The heads of these bolts fit into recesses counterbored into the base and table support so that they will not interfere with other project parts.

The oval holes for screws shown in the table supports will allow the top to expand and contract across the grain. If plywood is used, these oval holes are not required.

All the post corners should be rounded to a ¼-in. radius to soften the lines of the table.

If teak is used, the areas to be glued should be washed with lacquer thinner to remove resin before applying fresh, recently tested, casein glue.

Teak is abrasive in character and will dull noncarbide tools quickly.

Dining Table

Letter	Description	Number of pieces	Thickness	Width	Length
A	Top (edge-to-edge glued)	7	1-1/2	6	42
B	Table support	2	1-1/2	3	20
C	Post	4	2	3	24-1/4
D	Base	2	2	4	32

Notes
1. Material: teak or walnut.
2. Additional requirement: hex-head lag screws 1/4-in. x 3-1/2-in. (8), 1/4-in. x 2-1/2-in. (4).
3. Alternate heartwood on table top.
4. Cross-grain table supports, B, have oblong screw holes to allow expansion and contraction of top.
5. Posts, C, fit into 1/4-in.-deep dados in table supports, B.
6. Posts, C, fastened to table supports, B, and base, D, with lag screws screwed into ends of posts and into 1/2-in. anchor dowels fitted into inside edges of posts.
7. Table supports, B, joined with a cross-lap joint.

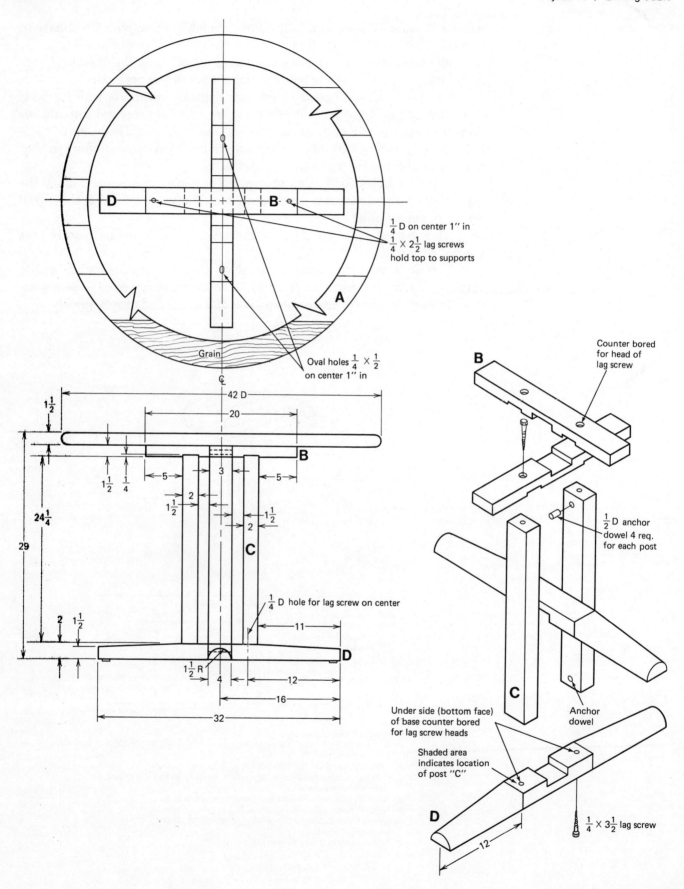

$\frac{1}{4}$ D on center 1'' in

$\frac{1}{4} \times 2\frac{1}{2}$ lag screws hold top to supports

Oval holes $\frac{1}{4} \times \frac{1}{2}$ on center 1'' in

Grain

A

B

D

C

42 D

20

$1\frac{1}{2}$

5

$1\frac{1}{2}$ $\frac{1}{4}$

3

5

B

$1\frac{1}{2}$

2

$1\frac{1}{2}$

2

C

$24\frac{1}{4}$

29

$\frac{1}{4}$ D hole for lag screw on center

2 $1\frac{1}{2}$

11

D

$1\frac{1}{2}$ R

4

12

16

32

Counter bored for head of lag screw

B

$\frac{1}{2}$ D anchor dowel 4 req. for each post

C

Anchor dowel

Under side (bottom face) of base counter bored for lag screw heads

Shaded area indicates location of post ''C''

D

$\frac{1}{4} \times 3\frac{1}{2}$ lag screw

12

PROJECT
12

Circular Bench or Table

This project can double as either a bench or a table. If desired, a circular foam cushion can be obtained to make the piece more comfortable when it is used as a bench.

The circular top of this piece is made up of eight 1-in. by 4-in. by 32-in.-long pieces. These pieces are joined edge to edge with dowel joints and with the heartwood on each piece alternated.

The leg rail support system is comprised of four legs and four rails. The rails are edge-lapped to form two crosses. The upper pair of rails serve to support the top and stabilize the upper ends of the legs. The lower pair of rails stabilize the lower ends of the rails.

The rail ends are shaped to form tenons which are 1 in. long, 3 in. wide, and ½ in. thick. (The rails are 3 in. wide.) In effect, they are similar to stub tenons often used in frame and panel construction. Since all four rails are identical, and all have these tenons on each end, all eight tenons can be cut at one time after the rails have been squared to final dimensions. This makes construction time considerably shorter.

The rail tenons fit into cutouts in the top and bottom ends of the legs. The cutouts are centered on the legs and measure ½ in. in width and 3 in. in length. Thus, when the legs and rails are assembled, an open mortise-and-tenon joint is formed. These cutouts can be made with a backsaw and chisels or they can be ripped on a table saw and bored out on a drill press. A chisel can be used for final trimming operations.

The rail edge lap joints are made by cutting a 1-in.-wide groove, 1½ in. deep into the edge of each rail at the centerline. Since these pieces are identical, the four rails can be ganged (fastened together), and the groove can be cut through all four rails at the same time. A miter box or table saw can be used to make these cuts. Several side-by-side cuts can be used to remove the bulk of the material between the two outside saw cuts. A chisel is used to complete the job.

The top is fastened to the rails with four counterbored wood screws. Two-inch No. 10s should be suitable. Note that on the cross-grain rail, oval counterbored holes are made for the screws to allow for cross-grain movement in the top.

If the table is to be used on an uncarpeted surface, four glides should be driven into the ends of the leg rail assemblies.

All joints should be glued. The tabletop is not glued to the rails.

Circular Bench/Table

Letter	Description	Number of pieces	Thickness	Width	Length
A	Top	8	1	4	32
B	Legs	4	1	3	17
C	Rails	4	1	3	30

Notes
1. Material for top: solid wood edge doweled, or veneered plywood or lumber cure.
2. Rails joined to legs with open mortise and tenon joints.
3. Top fastened to rails with screws counter bored through rails. In solid-wood construction screw holes in cross-grain rails must be oval for wood top movement.
4. Top doweled 2-in. from end and at 8-in. intervals with 1/2-in. dowels.

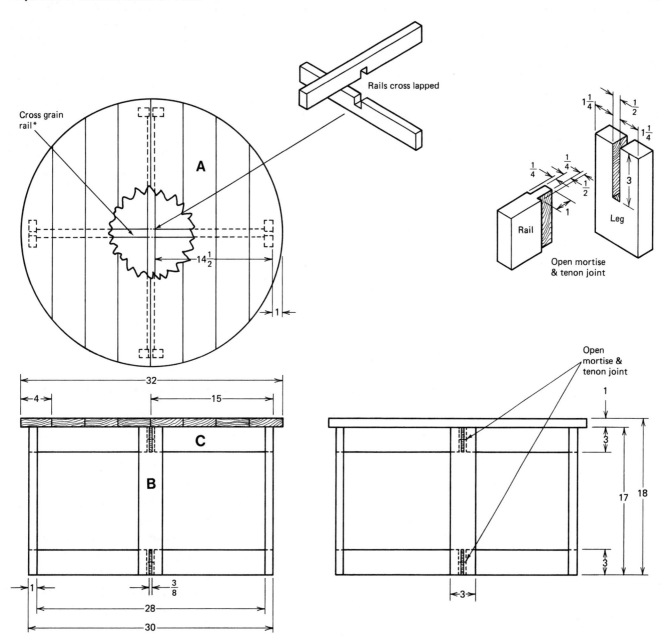

Rails cross lapped

Cross grain rail*

A

$14\frac{1}{2}$

1

$1\frac{1}{4}$ $\frac{1}{2}$

$1\frac{1}{4}$

$\frac{1}{4}$ $\frac{1}{4}$

$\frac{1}{2}$

3

1

Rail

Leg

Open mortise & tenon joint

32

4

15

C

B

1

$\frac{3}{8}$

28

30

Open mortise & tenon joint

1

3

17 18

3

3

PROJECT
13

Colonial Candle Table

The colonial candle holder was used to hold a candle at various heights to suit the needs of the user. It consists of a cross-bar base which supports an inverted table support. The circular table is mounted on a post edged with teeth. The teeth are engaged by a pawl which allows the table height to be changed. The bottom of the post is stabilized by a movable support bar whose tenoned ends ride in grooves cut into the edges of two vertical posts.

At first glance this project may seem somewhat complex. Further study of the drawing will reveal that the parts are fairly simple in design and their fabrication quite doable.

The base consists of two 1¾-in.-square pieces 13 in. long cross-lapped together at their respective center points. One of the base pieces is mortised in two places to receive the post, C, and the pawl post, D. The drawing indicates location and mortise size.

The two posts, C and D, can be cut from a single piece of 1⅛-in.-square stock. The length of each is 17 in., including the tenoned ends. The ends of each post are tenoned. The tenons are ⅜ in. thick and ¾ in. long. These can be cut at the same time. Even though the ends of part F are narrower than parts C and D, the tenons on the end of part F can be cut at the same time the other tenons are cut. Thus six tenons can be cut at one time.

Each of the posts, C and D, have a groove cut into their inside edge. This centered groove is ⅜ in. wide and ½ in. deep. This can be cut on a table saw using a dado cutter, or it can be cut with a power router fitted with a ⅜-in.-diameter bit. Of course, hand tools can also be used. A ⅜-in.-wide chisel would be the tool to use.

Note that part D is fitted with a pawl. The location of the pawl and its size and shape are given on the drawing.

The bottom ends of parts C and D fit into the mortise already cut in one of the base pieces. The upper ends fit into through mortises cut into the fixed bar, H. This ¾-in. by 1¾-in. by 10¼-in.-long bar is depicted in the drawing. In addition to the two through mortises for the top ends of the posts, it has a rectangular opening ¾ in. by 1 in. wide at its center to allow the movable post, F, to slide through it. The posts are fastened to part H with ¼-in.-diameter dowels.

The movable post, F, is cut from ¾-in. by 1-in. stock and is 14¾ in. long. As mentioned earlier, its ends are tenoned. The right edge of this post has a series of notches cut into it as shown in the drawing. These notches are made by making a series of saw cuts ¼ in. deep at ⅝-in. intervals. Then a chisel cut is made at an angle starting ⅜ in. below each of the saw cuts. A triangular-shaped cutout results.

The lower end of the movable post, F, fits into a mortise in the movable support bar, E. This bar is shown in an auxiliary top view between the two main views of the drawing. Its ends are tenoned to fit the grooves in the posts, C and D. It has a ⅜-in. by 1-in. mortise at its center to receive the bottom end of the movable post, F. The tenons on this piece should be trimmed to allow easy movement in the grooves of parts C and D.

The top, J, can be cut from a single piece of ¾-in.-thick stock so as to form a circle 10¼ in. in diameter. This tip is supported by part I, which is ¾ in. thick, 3 in. wide, and 8 in. long. It has a ⅜-in. by 1-in. mortise cut through it at its center to receive the top end of the movable post, F. Notice that it is fastened across the grain of the circular top with wood screws that fit into oblong holes. These holes allow the piece to expand and contract across the grain.

The base pieces can be made easy to disassemble if a suitable brass screw and wing nut are used to hold them together.

Colonial Candle Table

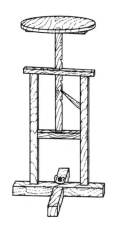

Letter	Description	Number of pieces	Thickness	Width	Length
A	Mortised base	1	1-3/4	1-3/4	13
B	Unmortised base	1	1-3/4	1-3/4	13
C	Post	1	1-1/8	1-1/8	17
D	Pawl post	1	1-1/8	1-1/8	17
E	Movable support bar	1	3/4	1-1/4	7
F	Movable post	1	3/4	1	14-3/4
G	Pawl	1	3/8	1/2	3-3/4
H	Fixed bar	1	3/4	1-3/4	10-1/4
I	Tabletop support	1	3/4	3	8
J	Tabletop	1	3/4	10-1/4	10-1/4

Notes
1. Material: oak or pine.
2. Posts 15-1/2 in. long + two 3/4-in. long tenons = 17 in.
3. Movable support bar 6-in. long + two 1/2-in.-long tenons = 7 in.
4. Upper corners and end corners, A and B rounded to 1/4-radius.
5. Parts A and B may be fastened with a brass bolt and wing nut 3/8-in. x 2 in. bolt.
6. All dowels shown 1/4 in. in diameter.
7. Base parts A and B held together with brass wing-nut and screw.

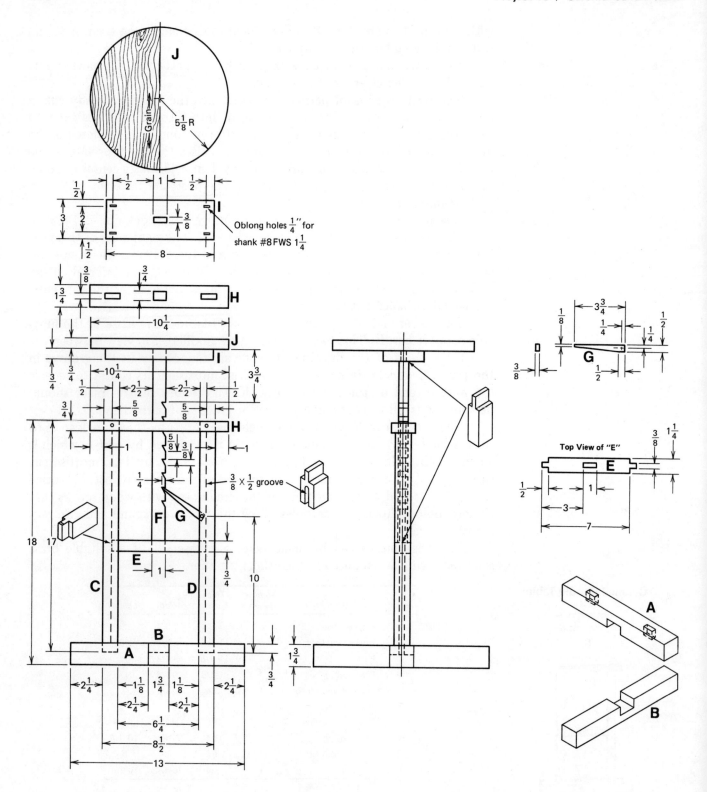

PROJECT
14

Valet Seat

This piece is a chair which is designed to do the job a regular chair is often expected to do. It is a chair that will hold a shirt or jacket and a pair of trousers and a person while he or she dresses or undresses.

The chair has only three legs, which simplifies its construction, and its back sports a hanger and a trouser bar. If desired, the hanger-shaped component shown in the drawing can be replaced with a well-made wooden hanger that has been cut out to take the place of part F.

The front legs, A, are fashioned from $1\frac{1}{2}$-in. by 2-in stock. Each is 17 in. in length. The legs taper in two directions to 1 in. square. The upper $2\frac{1}{2}$ in. of the leg are left rectangular in cross section. If desired, the taper on the legs can be cut on a wood lathe, resulting in rounded legs. Each of the legs is fastened to the front rail, C, with plugged screws. Notice that these screws are driven into $\frac{1}{2}$-in.-diameter anchor dowels fitted into the lower edge of the front rail.

The front rail, C, is a piece measuring $1\frac{1}{2}$ in. by $2\frac{1}{2}$ in. by 14 in. in length. Its inner face is dadoed with a $\frac{1}{2}$-in. by $1\frac{1}{2}$-in. cut made as shown in the detailed view at the far-right side of the drawing. This dado receives the end of the center rail, D.

Part D is $1\frac{1}{2}$ in. thick, $2\frac{1}{2}$ in wide, and $12\frac{1}{2}$ in. long. Its back end is shaped into a $\frac{1}{2}$-in.-wide tenon. This tenon fits into a mortise centered on the leg back support, B, as shown in the drawing.

The leg back support, B, can be cut from a piece of wood 2 in. thick,

6 in. wide, and 36½ in. long. The layout of the piece is shown on the drawing. Notice that lines are drawn along this 3-in.-wide piece at 2-in. intervals. These lines are square (90°) to the edge of the piece. Each line is drawn to the length indicated. For example, the first line is 2⅞ in. long, the next 3 in., the next 3⅛ in., and so on. When all the lines are drawn, a curve is drawn connecting the ends of the lines. This will result in a reproduction of the curve. The inside lines are straight.

The seat, E, can be made from a piece of ¾-in.-thick cabinet hardwood plywood or it can be made up of smaller pieces joined edge to edge. If the cabinet plywood is used, the ends can be rounded over and finished. Banding is not necessary. The seat is fastened to the front and center rails with counterbored wood screws driven through the rails and into the seat. The seat ties the rails together and stabilizes the whole piece. For added rigidity, dowels can be inserted through the top and into the legs.

The hanger, F, is laid out in a way similar to that used for the leg back support. A detailed drawing of the part is given at the upper left of the drawing. Notice that the hanger is made in two matching halves that cross-lap the post.

The top bar is a 1-in.-diameter dowel 16 in. long.

Valet Seat

Letter	Description	Number of pieces	Thick-ness	Width	Length
A	Leg, front	2	1-1/2	2	17
B	Leg, back support	1	2	6	36-1/2
C	Front rail	1	1-1/2	2-1/2	14
D	Center rail	1	1-1/2	2-1/2	12-1/2
E	Seat	1	3/4	18-1/2	15
F	Hanger	2	1	6	8-3/4
G	Pants bar	1	1	1	16
H	Leg support bar	1	3/4	3/4	15

Notes
1. Material: maple or oak.
2. Approximate width of piece for layout given for part B.
3. Dowel.
4. Parts B and D joined with A mortise-and-tenon joint: 1/2-in.-wide tenon 1-in.-deep mortise.
5. B and G mortise-and-tenon joint: 3/8-in.-wide tenon 3/8-in.-deep mortise.
6. D and C joined with dado joint 1/2 in. deep and 1-1/2 in. wide.
7. Legs.
8. Seat fastened to rails with 1-1/2-in. No. 10 flathead wood screws 1 in. in from ends of seat centered on rails. Plugged screws in front rails centered in oblong slots to allow seat to expand and shrink across the grain.
9. Hangers 1-1/2 in. lapped into part B, as shown. Part F glued and doweled with 3/8-in. through dowel.
10. Legs, A, and lower section of part B tapered all corners rounded 1/4-in. in radius.
11. Part H penetrates legs, A, to a depth of 1/2-in.
12. Seat, E, make up of 4-in.-maximum width pieces joined edge to edge with dowels 3/8-in. in diameter.

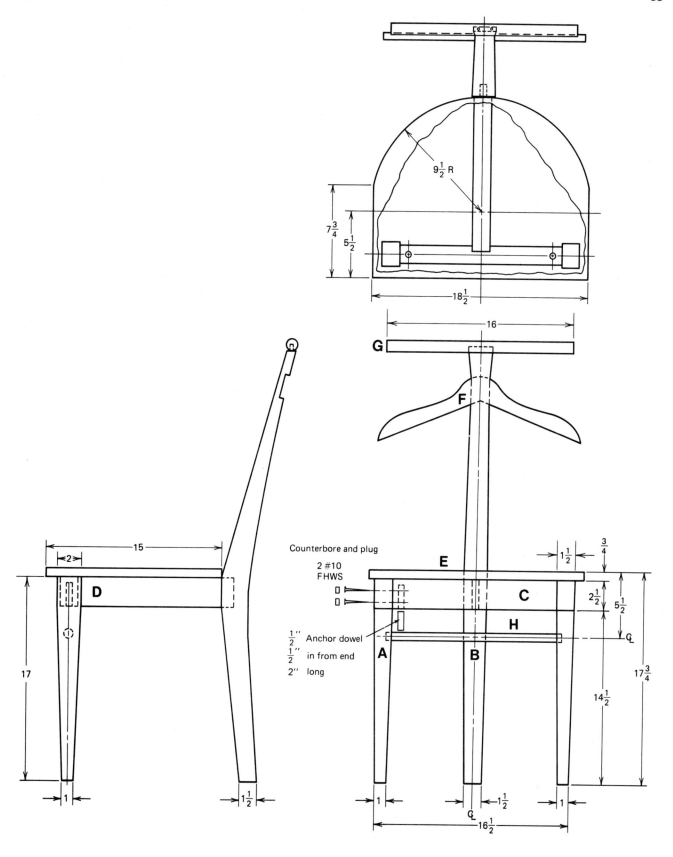

$9\frac{1}{2}$ R

$7\frac{3}{4}$

$5\frac{1}{2}$

$18\frac{1}{2}$

16

G

F

15

2

D

E

C

H

$1\frac{1}{2}$

$\frac{3}{4}$

$2\frac{1}{2}$

$5\frac{1}{2}$

$17\frac{3}{4}$

$14\frac{1}{2}$

Counterbore and plug

2 #10
F H W S

$\frac{1}{2}$" Anchor dowel
$\frac{1}{2}$" in from end
2" long

A

B

17

1

$1\frac{1}{2}$

1

1

$1\frac{1}{2}$

$16\frac{1}{2}$

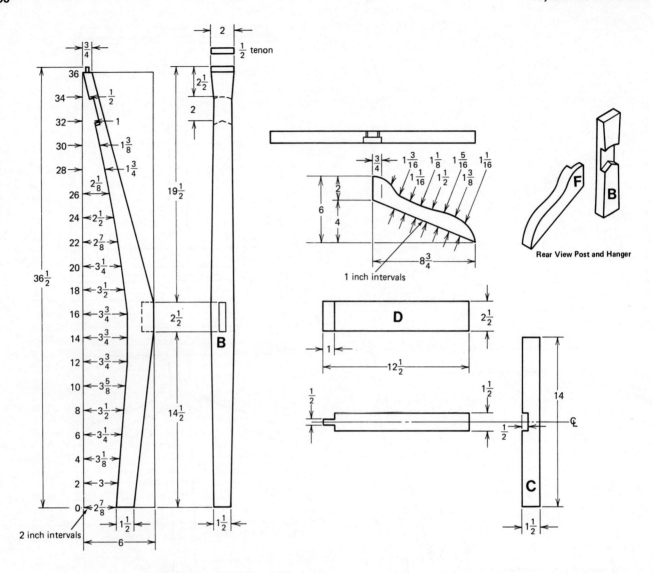

Rear View Post and Hanger

PROJECT
15

Tea Cart

This project is a useful addition to any dining room or kitchen. It serves as a trolley to carry several serving platters of food from the kitchen to the dining area. It also provides extra table space at the dining table. It can also be used as a portable bar or as a small buffet table.

The design shown here is very simple and is therefore easy to construct. More advanced craftsmen might want to sculpt parts or make other changes. Any changes will be cosmetic and will not materially add to the usability of the cart.

This piece is made up of three sets of identical parts. Since identical parts can be fashioned quickly by making multiple cuts, or by ganging pieces and treating them as a single piece, considerable time can be saved in construction.

Part A is used as the leg piece. Four are required. Each is 1 in. thick, 2 in. wide, and 28 in. long.

End rails, B, are also cut from 1 × 2s 17 in. long. Four are needed.

Side rails, C, are 1 × 2s 27 in. long. Four are required.

All of these parts, plus the handle, can be cut from approximately 30 ft of 1-in. by 2-in. material. Note that in this case the 1-in. and 2-in. dimensions are finished dimensions, not nominal sizes.

All the side rails, C, have a ½-in.-wide groove, ¼ in. deep cut into their inner faces. This is shown clearly in the isometric view on the drawing. These grooves can be cut on a table saw using a dado cutter or with a power router. The grooves receive the top and bottom shelf sides.

The ends of each of the shelves are supported by a wooden strip, G, which is glued onto the end rails, B. This is shown in several places on the drawing.

In assembling the cart, the ends should be assembled first. The ends are made up of two legs, A, and two end rails, B. These pieces are held together by ½-in.-diameter dowels which are used in blind dowel joints. If desired, through dowel joints or plugged screws can be used instead. It is important to note that if through dowels or screws are used, care must be taken to make sure that these screws or dowels do not interfere with the dowels used to connect the side rails to the legs. See the drawing for detailed information.

The assembled cart ends are joined to the legs with ½-in.-diameter through dowels as shown in the drawing. After one rail is in place, the shelves must be placed in the rail grooves. When the opposite rail is installed, the other edge of the shelf is placed in the groove in the second rail.

The handle is cut from a piece of ¾-in. by 1½-in. stock. It is 14 in. long. An option would be to replace this handle with a lathe-turned handle 1 in. in diameter. The handle support pieces, F, can be cut from 1-in.-diameter dowels. The handles are attached to the end rail of the cart with screws driven through the end rails and handle supports and into the handle. A second identical handle may be added to the other rail end if desired.

The cart is made mobile by attaching four ball casters to the legs.

Some serving carts have large wheels attached to two of the legs in place of the casters.

Tea Cart

Letter	Description	Number of pieces	Thickness	Width	Length
A	Leg	4	1	2	28
B	End rail	4	1	2	17
C	Side rail	4	1	2	27
D	Shelf	2	1/2	19	27
E	Handle	1	3/4	1-1/2	14
F	Handle support	2	1	1	1
G	Shelf support strip	4	1/4	1/2	18-1/2

Notes
1. Material: oak, walnut, or cherry. For the shelves, veneered (oak, walnut, cherry) plywood or Formica-covered fir plywood. If Formica is used, increase groove width in part C to 9/16-in. x 1/4-in. deep.
2. Shelf support strip, G, can be glued to (end rails), B, or a groove can be cut into part B when the groove is cut into part C. Then insert part G, which must be 1/2 in. x 1/2 in. instead of 1/4 in. x 1/2.
3. Hooded rubber ball casters with brass finish that raise legs approximately 2-1/2 in. are required to complete the cart.
4. All outside corners rounded to 1/4-in. radius after end rail leg units are assembled.

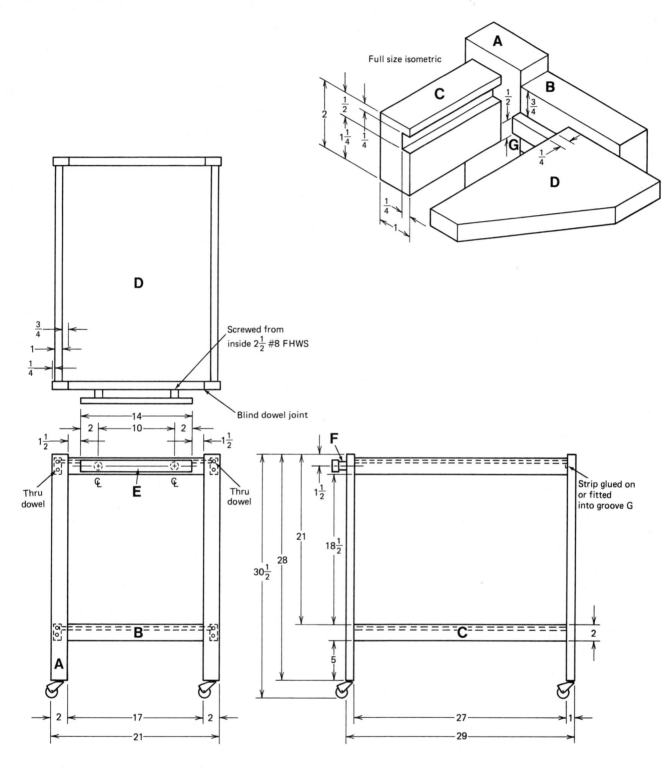

Full size isometric

Screwed from
inside 2½ #8 FHWS

Blind dowel joint

Thru
dowel

Thru
dowel

Strip glued on
or fitted
into groove G

PROJECT
16

Knock-Down Drafting Table

This fold-up drafting table is ideal for the kind of occasional use that most people make of a drafting table. This table, when set up, has a front-hinged top which can be tilted and set at a variety of angles. When taken down, the table folds up into a compact unit which can easily be stored for future use.

A standard-size wooden drafting board can be used in place of the wooden board described here. This will save considerable construction time. Of course, the dimensions of the table supporting assembly will have to be changed to fit the drawing board selected.

The table support assembly consists of three rectangular frames. The largest frame forms the back of the assembly while the two smaller units form the sides. These side assemblies are designed to fold flat against the back. The three frames are all constructed of ¾-in. by 3-in. stock joined at right angles using half-lapped joints. Since these joints are identical, they can easily be fabricated once the initial setup is made. The lapped joints can be made on a radial arm saw, with dado cutters to speed things up. A table saw can be used, or a power router used in conjunction with a clamp-on T-square router base guide will also do the job. This T-square is a shop-built device.

The top frame is also made up of ¾-in. by 3-in. stock half-lapped at the corners. These pieces can also be cut using the same setup used for the table support assembly frames.

The drawing board is made from a single piece of ¾-in.-thick plywood, edges banded, which is attached to the top frame with a piano hinge. A drawing board support, H, is described in detail in the drawing. If desired, a suitable hardware device can be substituted for the table support shown.

The tabletop top frame mounts on dowels fixed to the upper rails of the base back and sides. An alternative method of fastening is to install carriage bolts through the side and back rails so that they project above the frame edges about 1 in. These bolts would fit into matching holes bored through the top frame. Wing nuts and washers would hold the assembly together.

Notice the arrangement of the sides and base as shown in the side view located at the bottom right of the drawing. A leg spacer, I, is installed to position the right side of the base so that it can be folded flat against the back of the base after the left side has been folded.

The inside and outside corners of the frames are rounded over using a rounding-over bit in a power router after the frames are assembled.

Knock-Down Drafting Table

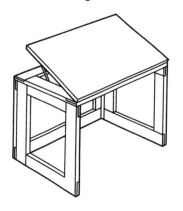

Letter	Description	Number of pieces	Thick- ness	Width	Length
A	Base stile (vertical)	6	3/4	3	26-1/2
B	"Short" side rails (base)	2	3/4	3	21-1/2
C	"Long" side rails (base)	2	3/4	3	22-1/4
D	Back rails (base)	2	3/4	3	31
E	Drawing board	1	3/4	31	23
F	Frame rail top	2	3/4	3	31
G	Frame stile top	2	3/4	3	23
H	Drawing board support	2	1/2	3/4	14
I	Leg spacer	1	3/4	3/4	26-1/2

Notes
1. Material: frame, pine or maple; drawing board, 3/4-in. A through C plywood; top support, maple or other hardwood.
2. Back rails, D, same dimensions as top frame rails, F.
3. Top frame secured to base by dowels.
4. All rails and stiles 1/2-in. lapped joinery.
5. Drawing board edges covered 1/4-in. x 3/4-in. hardwood.

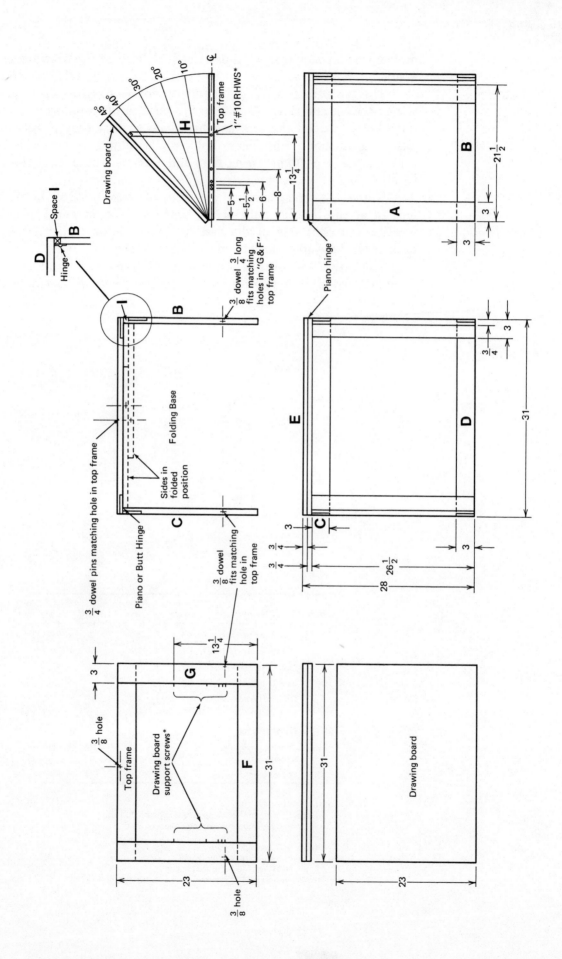

Drawing board

Top frame

1" #10 RHWS*

H

45° 40° 30° 20° 10° ℄

13¼

8
6
5½
5

3/8 dowel 3/4 long fits matching holes in "G & F" top frame

Space I

D

B

Hinge

I

B

3/4 dowel pins matching hole in top frame

Sides in folded position

Folding Base

Piano or Butt Hinge

C

3/8 dowel fits matching hole in top frame

A

B

21½

3

3

Piano hinge

E

D

3

¾

31

C

3
¾
¾

26½

28

Top frame

3/8 hole

3

G

13¼

Drawing board support screws*

F

31

3/8 hole

23

Drawing board

31

23

62

PROJECT
17

Parson's-Style Table

Parson's-style tables have the advantage of fitting in with almost any other style of furniture. The table's straight lines and simple form make it unobtrusive.

The version presented here has overall dimensions of 14½ in. by 59¼ in. by 27 in. high. Such a table would be suitable for use against a wall in an entrance area or in place behind a sofa. Other overall dimensions for tables are given at the end of this description. Keep in mind that as the table increases in overall size, its top and leg thickness should also increase.

The tabletop in this piece is a piece of ¾-in.-thick plywood measuring 13¾ in. by 58¼ in. The upper face of the plywood should be B grade minimum. This piece of plywood fits into rabbets cut into the side and end rails of the table.

The side and end rails can be cut from a single piece of ¾-in. by 2¼-in. material 12 ft 2 in. long. The ½-in.-wide, ¾-in.-deep rabbet can be cut into a 10-ft-long piece of this rail material from which the side rails, B, will be cut. The rabbets in the side rails, A, are stopped rabbets and have to be cut after the rails are cut to their finished 60-in. length. The enlarged-scale isometric drawing showing the relationship of the rails illustrates the stop rabbet.

The rails are supported by a shoulder cut into the upper end of each leg. The exploded isometric view shows this shoulder. The assembly isometric view shows the pieces after assembly. The rails should be glued and screwed to the legs and the leg braces. All screw heads must be covered with wooden plugs and sanded flush.

One structural problem is presented by this type of table design. Since the legs have very little support (no lower rails, for example), they can be damaged easily if the table is pushed while its legs remain "stuck" in carpeting. Two steps are taken in this design to strengthen the leg joints. First, the legs are made of strong hardwood such as maple. Second, the two leg braces are attached to each leg and to the underside of the table. This supporting system of braces is shown clearly in the isometric views.

When the table is assembled and glued up, the lower ends of the legs should be tacked to temporary braces running from leg to leg. The braces will help in keeping the legs square to the tabletop while the glue cures. Similar braces should be installed whenever the table is shipped or stored away.

After assembly and glue curing the table should be sanded flat and smooth. Medium or 100-grit abrasive is suitable for this operation. Dusting with a tack cloth is the last step prior to the installation of the plastic laminate.

Many tables of this type are covered with a plastic laminate (such as Formica® brand). This material is heat and mar resistant, and its use eliminates the need for finishing. Most serious craftsmen select from among the many solid colors and finishes that are available. Wood-grained material is generally considered to represent less than an honest use of the material. Wood should look like wood and plastic like plastic. A recently available type of plastic laminate called Colorcore® brand* is the same color all the way through. Standard laminates have a dark brown core behind the colored face. Colorcore® is more expensive than standard plastic laminates, but it makes plastic laminate edges almost invisible. The procedure for laminate installation given here is designed to reduce edge visibility. The inner surfaces of the legs should be covered first. These pieces cannot be trimmed in place after adhering as is the usual practice because the table rails will interfere with the router base during trimming. They are trimmed by clamping the rough-cut plastic laminate to a piece of wood that is the same width as the area of the leg to be covered. It is a good idea to make an extra leg for this purpose. After trimming, the strip of plastic laminate is carefully placed on the leg in its final position. Two coats of contact cement are applied to all wooden and plastic laminate surfaces that are to come into contact with each other. Follow label directions on the contact cement carefully. Solvent-based cements generally work better than the latex versions.

After all eight inside surfaces of the legs are done, the short rails are tackled next. These pieces must be mitered. They are cut by clamping the rough-cut pieces of plastic laminate to a piece of wood that is mitered on each end and is the same length as the side of the tabletop. The plastic laminate should be at least ½ in. wider than the rail. The plastic laminate is coated with cement and adhered to the cement-covered end of the table. The miters on the plastic laminate must line up with the miter lines laid out on the table end. Now the pieces of plastic laminate needed to cover the ad-

*Formica® brand and Colorcore® brand plastic laminates are registered trademarks of the Formica Corporation.

joining surface of each of the legs are prepared. These pieces will be over-sized in length and width, but they will have a miter cut on one end. After coating the required surfaces with contact cement, the mitered edges of the plastic laminate will be carefully placed against the mitered edge of the plastic laminate already adhered to the rail. Then the rest of the piece will be lowered into position. Once the ends of the table are covered, the plastic laminate will be trimmed flush with a router fitted with a flush-trimming plastic laminate bit.

The long rails and leg fronts are treated in a similar fashion. The last surface covered is the top. As each piece of plastic laminate is laid down, it should be pressed against the table with a hammer and wooden block or with a J roller made for that purpose.

Excess contact cement can be removed with lacquer thinner.

One way to reduce the visibility of any gaps in plastic laminate joints is to paint the area under the joints with a paint that is close in color to the color of the plastic laminate used.

Alternate Dimensions
for Parson's-Style Tables (in.)

Length	Width	Height
18	18	17
20	16	18
24	24	16
30	24	23
42	23	24
48	24	30

Parsons-Style Table

Letter	Description	Number of pieces	Thick-ness	Width	Length
A	End rail	2	3/4	2-1/4	13
B	Side rail	2	3/4	2-1/4	60
C	Leg brace 1	2	3/4	1-1/2	10
D	Leg brace 2	2	3/4	1-1/2	13
E	Leg	4	2-1/4	2-1/4	26-1/4
F	Top	1	3/4	13-3/4	59-1/4

Notes
1. Materials: legs/rails/braces, wood—choice of maker.
2. All parts fastened with glue and flathead wood screws.
3. End and side rails fastened to top of leg with screws from the inside.
4. Leg brace 1 screwed through its edge into the top.
5. Leg brace 2 screwed through its edge into the top and face-screwed to the top of the leg.
6. Plywood top glued into rail rabbets; rabbets in rails 1/2 in. wide by 3/4 in. deep.
7. Table covered with plastic laminate (Formica) with vertical surfaces of legs and rails mitered.

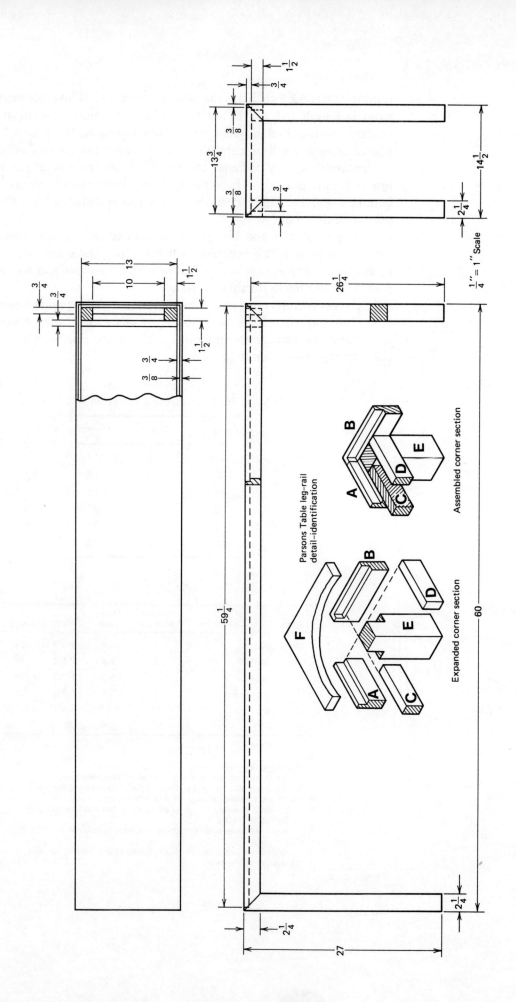

Parsons Table leg–rail
detail–identification

Assembled corner section

Expanded corner section

$\frac{1}{4}'' = 1''$ Scale

PROJECT
18

Knock-Down Cradle

Cradles are used for a relatively short time because the rapidly growing child soon outgrows it. It is a real plus to have a cradle that can be disassembled or knocked down for storage. This cradle can easily be disassembled into the crib, D to G, the vertical support assembly, A and B, and the rail, C.

This cradle is colonial or Early American in character. It can be finished with white paint, leadless, of course, or it can be given a transparent finish such as varnish or lacquer.

The cradle assembly is made up of parts D to F. The crib end, F, is composed of three pieces which are ¾ in. thick. The two outer pieces are 7 in. wide and the inner piece is 4 in. wide. (See the detailed end view of the cradle on the upper left side of the drawing.) The 4-in.-wide center piece has the same shape as the narrow end of the vertical support, B. The vertical support should be made first and then used as a template to lay out the center piece in part F. The tapered part of the center piece is easier to cut before the three pieces making up the end are assembled. These three pieces should be assembled using ⅜-in. dowels as described in the information given on edge-to-edge dowel joints. Since two end pieces, F, are required, the pieces should be fastened together with double-sided carpet tape and treated as a single piece. The tapers can be cut by hand or on a table saw using a taper jig.

The cradle (crib side) side, E, is cut from ¾-in. stock and is 11¹¹/₁₆ in. wide and 38 in. long. Note that the edges of the side are not square (90°) to the face. The edge is at an approximately 80° angle. This angle can be picked off the drawing with a T bevel and can be used to set the table saw blade. If hand tools are used, the T bevel is used in place of the try square during planing.

The crib bottom, D, is cut from a piece of ½-in.-thick exterior plywood. It is edged with ½-in. quarter-round molding. The crib can be assembled using plugged screws or through dowels.

The cradle stand is made up of a support assembly consisting of the vertical support, B, the base, A, and a rail, C. Parts A and B can be fastened together with 1½-in.-long No. 10 flathead screws driven through and flush with the base, A, and into the end of the vertical support, B.

The scroll-sawed design on the crib sides, E, should be cut while the sides are fastened together with double-sided carpet tape.

The rail, C, is 1 in. thick, 6 in. wide, and 45 in. long. Its ends are narrowed to 4-in.-wide projections 2½ in. long. (See the detailed view of part C at the top right of the drawing.) These projections are tenons that fit through mortises (rectangular holes) cut through the vertical support, B. Tapered pins, I, are driven into holes in part C. These pins pull the vertical support, B, up tightly against the shoulders of the rail, C. This is how the tusked tenon joint works. Knocking the pins out allows easy disassembly.

It is important to note that the narrow ends of the vertical support, B, and the matching ends of the crib end, F, are reinforced with slip feathers which are glued into a groove in the end of each piece. (See the detail at the top left side of the drawing.)

The construction of the turned pins, G, allows the pins to be disassembled by removing a single screw. (See the detail at the top right of the drawing.)

When the cradle is assembled, the swing of the crib should be limited by the upper edge of the rail, C, which comes into contact with the edge of the crib bottom when it swings too far.

Knock-Down Cradle

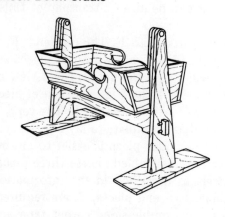

Letter	Description	Number of pieces	Thickness	Width	Length
A	Base	2	1	6	24
B	Vertical support	2	1	8	39
C	Rail	1	1	6	45
D	Crib bottom	1	1/2	15	38
E	Crib side	2	3/4	11–11/16	38
F	Crib end*	2	3/4	18	23–1/2
G	Maple–turned pin	2	1–1/2	1–1/2	3–1/2
I	Pins—turned tenon	4	3/4	3/4	3

Notes
1. Material: pine, with plywood crib bottom.
2. Crib bottom 1/2-in. plywood screwed into crib sides and end, 1/2-in. quarter-round molding applied to ends and edges of plywood.
3. Pin, G, fastened to part B with 1/4-in. dowel. Removable cap, H, makes crib removable.
4. Pins for tusked tenon fashioned from 3/4-in. birch dowels. Make cradle "knock-down." Pins cut tapered lengthwise for tight fit.
5. Crib sides, E, screwed to crib ends, F, with 1-1/2-in. No. 8 flathead wood screws. Holes plugged to cover heads.
6. Crib end, F, fabricated from three pieces doweled edge to edge, two are 7" wide, 11-1/2" long, one is 4" wide and 23-1/2" long.
7. All outside corners rounded 1/4 in. in radius.

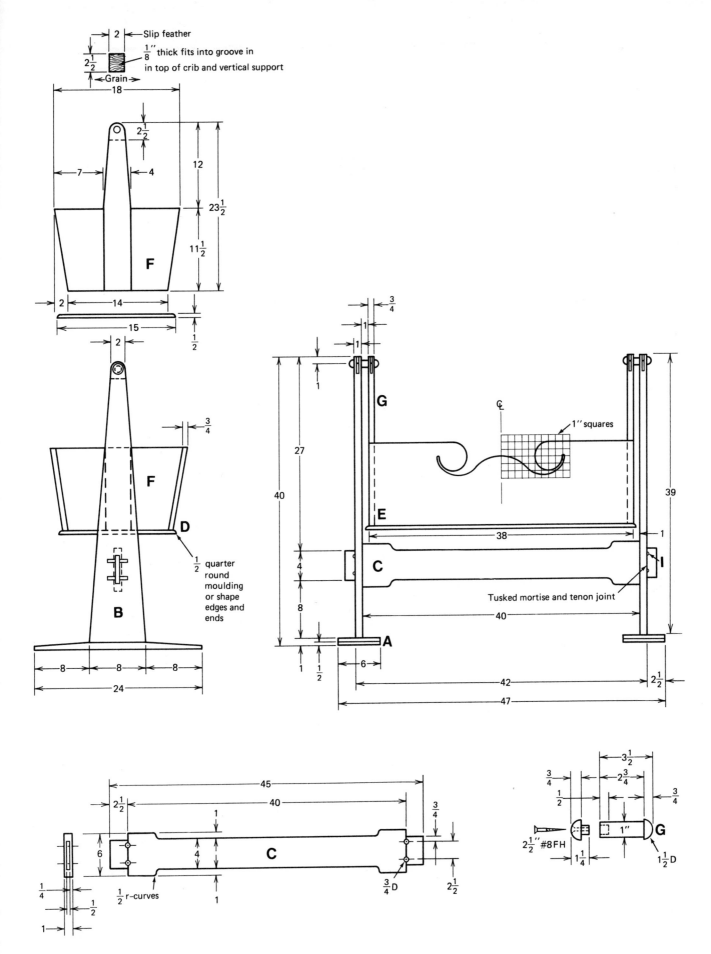

2 ─ Slip feather
$\frac{1}{8}''$ thick fits into groove in
in top of crib and vertical support
$2\frac{1}{2}$
←Grain→
18

$2\frac{1}{2}$
7 ─ 4
12
$23\frac{1}{2}$
$11\frac{1}{2}$
F
2 ─ 14
15
$\frac{1}{2}$
2

$\frac{3}{4}$
F
D
$\frac{1}{2}$ quarter round moulding or shape edges and ends
B
8 ─ 8 ─ 8
24

$\frac{3}{4}$
1
1
G
27
40
E
$\frac{1}{C}$
1'' squares
38
C
I
Tusked mortise and tenon joint
40
4
8
A
1
$\frac{1}{2}$
6
42
$2\frac{1}{2}$
47
39

45
40
$2\frac{1}{2}$
1
$\frac{3}{4}$
6
4
C
$\frac{3}{4}$D
$2\frac{1}{2}$
$\frac{1}{4}$
$\frac{1}{2}$
1
$\frac{1}{2}$ r-curves
1

$3\frac{1}{2}$
$2\frac{3}{4}$
$\frac{3}{4}$
$\frac{1}{2}$
$\frac{3}{4}$
1''
G
$2\frac{1}{2}''$ #8FH
$1\frac{1}{4}$
$1\frac{1}{2}$D

PROJECT
19

Hall Cabinet

This project was designed and constructed by Dean A. Van DeCarr. The original purpose of the cabinet was to store rubber boots. Since these boots were often wet, the bottom of the cabinet was fitted with a removable metal box. In addition, the back of the cabinet was left open to allow the circulating air to dry out the boots. The version of the cabinet shown here provides storage in a drawer and in a large space under the drawer. The piece is Early American in style. The original was made out of walnut, but other woods can be substituted.

This cabinet utilizes frame-and-panel construction in its doors and sides. Frame-and-panel construction allows the panels to expand and contract across the grain as humidity levels change in the room. For this to happen, the panels must fit in the frames loosely enough to permit the movement. In addition, the frames must be slightly smaller than the grooves that frame them. In general, wood can be expected to expand or shrink about ⅛ in. for each 12 in. of width. The length of a piece of wood is not affected in a significant way as humidity changes. Any thick finish applied to frames and panels will crack as the expansion and contraction takes place, so thin finishes such as oil are ideal.

This cabinet is built on a base or plinth. The plinth is made up of a front, A, two sides, W, and a back, K. The front is joined to the sides with a miter joint, which eliminates end grain. The back fits inside the sides. The top view of the plinth, at the lower right of the drawing, shows the arrangement of the pieces. Note that corner blocks are used to reinforce the corner joints. This drawing also gives a cross-sectional view of the front and sides. It shows the shaped upper edge and the rabbet on the back face of the piece. The ½-in.-thick plywood bottom fits into the recess formed by the side and front rabbets and stiffens the whole plinth.

The body or carcass of the cabinet is mounted on the foundation formed by the plinth. Notice that it is constructed around four corner posts. These can be seen in cross section in the top view of the carcass with the top removed.

The posts have grooves cut into them which receive tenons cut into the rails on the sides, T, U, and V, and the top and rabbets on the door frames, B and O. The door frame top, M, also fits into these post grooves. The back posts have a rabbet on the back face in place of a second groove. This can be seen in the top view of the carcass with the top removed.

The carcass sides, which fit into the post grooves, consist of top rails, T, raised panels, S, center rail, U, and bottom rail, V. An exploded view of the post, panel, and rails is shown at the right center of the drawing. The raised panels are 11¾ in. by 11⅝ in. (part S). They are fabricated from several narrow pieces of ¾-in.-thick material, joined edge to edge with dowels. These "square" panels can be shaped on a table saw or with a shaper, if suitable cutters are available. A cross-sectional view of the panel edge can be found in the exploded view at the right center of the drawing.

These panels are *not glued* into their surrounding frames. The frames *are glued* into the posts.

The front doors have a serpentine-shaped upper rail. This curve can be laid out by reproducing the 1-in. squares full sized on the layout. A full-sized template of this curve should be made and used to lay out all of the curved parts.

The door frames and panels are made in much the same way as the square panels, S, and their door frames were made. A table saw can be used for the straight parts of the door but a shaper is required for the curved section. Some power routers may also be capable of doing the job. Of course, a skilled woodworker could cut the curved tapered sections by hand.

The cabinet top, P, should be built up using three 3⅞-in.-wide pieces joined edge to edge, with heartwood reversed to limit warping. The edges and ends of the top can be shaped with a router. The top should be secured to the carcass with steel S-clips. No glue should be used. This will allow the top to expand and contract across the grain. Veneered plywood is another option in place of the glued-up material. This would have to be framed with matching hardwood to allow shaping and to cover the edges of the plywood.

Detailed information on drawer construction is given in the top and sectional views of the drawer provided at the upper right side of the drawing.

Hall Cabinet

Letter	Description	Number of pieces	Thickness	Width	Length
A	Plinth front	1	3/4	4	37
B	Door frame bottom	1	3/4	2	34
C_L	Left door rail bottom	1	3/4	1-1/2	15-3/4
C_R	Right door rail bottom	1	3/4	1-1/2	15-3/4
D	Left door panel	1	3/4	13-3/4	16
E	Right door panel	1	3/4	13-3/4	16
F_L	Left door left stile	1	3/4	1-1/2	13.5
F_R	Right door right stile	1	3/4	1-1/2	13.5
G_L	Left door top stile	1	3/4	4	15-3/4
G_R	Right door top stile	1	3/4	4	15-3/4
H	Center divider	1	3/4	1-1/2	19-1/2
J_L	Left door right stile	1	3/4	1-1/2	18
J_R	Right door left stile	1	3/4	1-1/2	18
1K	Base back	1	3/4	3-1/2	35-1/2
L	Post	4	1-1/2	2	27-1/4
M	Door frame top	1	3/4	5	34
N	Drawer front	1	3/4	5-3/4	33-1/2
O	Top door frame	1	3/4	2	34
P	Top cabinet	1	3/4	15-1/2	37
Q	Drawer rail/corner block	4	3/4	3/4	14
R	Bead 1/2 round edge	2	1/4	2	77-1/2″
S	Side panels	4	3/4	11-3/4	11-5/8
T	Top side rail	2	3/4	2	12
U	Center side rail	2	3/4	1-1/2	12
V	Bottom side rail	2	3/4	2	12
W	Plinth side	2	3/4	4	15-1/2
X	Plinth corner block	4	2	2	3-1/2
Y	Bottom plinth (plywood)	1	1/2	14-7/8	36-1/4
AA	Drawer front	1	3/4	5-3/4	33-1/2
BB	Drawer side	2	1/2	5-3/4	14
CC	Drawer bottom (plywood)	1	1/4	13-3/4	33

Notes
1. See detail for door construction and top door frame. Door panel dimensions given for square cut stock. All panels unglued to their rails and stiles.
2. See detailed drawing for drawer construction.
3. Drawer rails top screwed to top and to part T. Similar pieces used at bottom screwed into base (plinth) and part V.
4. ½-round bead fitted to drawer opening and installed with mitered corners.

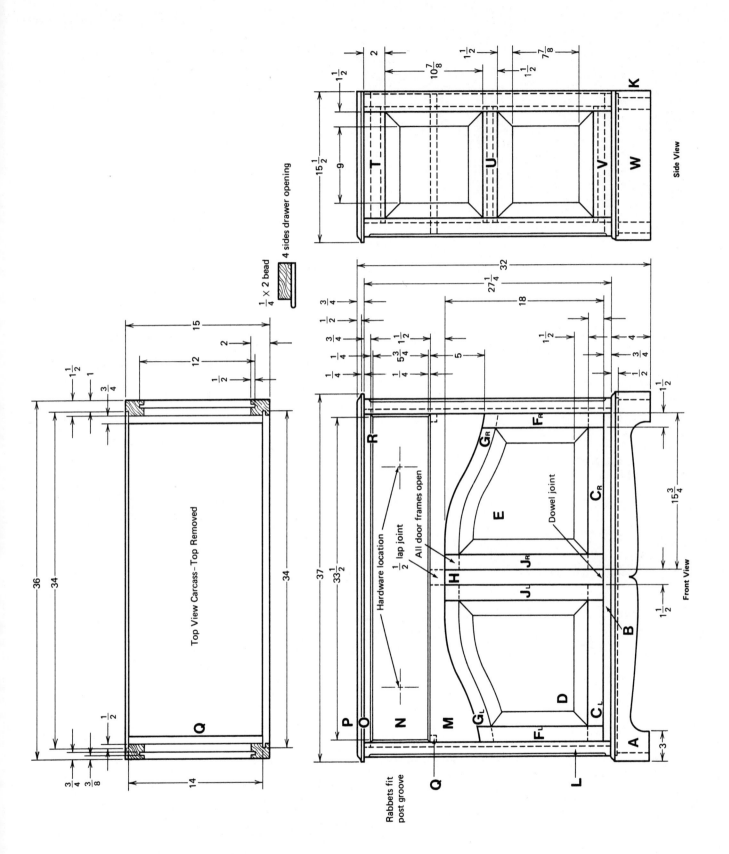

Top View Carcass–Top Removed

Side View

Front View

4 sides drawer opening

$\frac{1}{4} \times 2$ bead

Hardware location

$\frac{1}{2}$ lap joint

All door frames open

Dowel joint

Rabbets fit
post groove

73

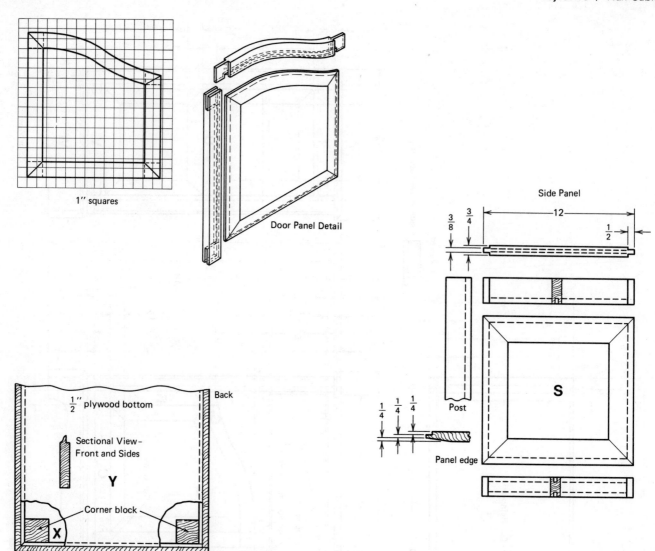

1″ squares

Door Panel Detail

Side Panel

12

Post

Panel edge

S

½″ plywood bottom

Back

Sectional View–
Front and Sides

Y

Corner block

X

Top View of Plinth (Base)

Drawer Details

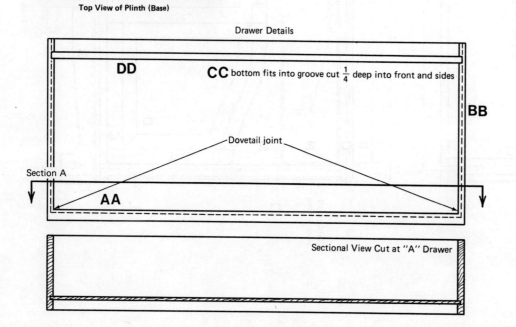

DD   **CC** bottom fits into groove cut ¼ deep into front and sides

BB

Dovetail joint

Section A

AA

Sectional View Cut at "A" Drawer

PROJECT
20

Seven-Drawer Chest

This project was designed, drawn, and built by Raymond Langer. This is a lingerie chest. Its seven drawers provide garments for each of the seven days of the week. Because the cabinet is relatively narrow, 24 in., it can often be placed against an otherwise unusable section of the wall.

This design utilizes many dovetail joints for maximum strength. It also includes dust panels which support the drawers and keep dust from falling from one drawer to the next as the drawers are used.

Cherry or walnut would be appropriate woods for the piece. The style suggests Early American design. This piece was constructed exclusively of a solid hardwood. However, the sides and top could be made of hardwood-veneered plywood with banded and framed edges and ends.

The base of this cabinet consists of a cabinet bottom, O, measuring ¾ in. by 15½ in. by 24 in. It is supported by cabriole legs. The bottom, O, is made up of four narrow pieces joined edge to edge with dowels. The front legs are cut from solid glued-up blocks of wood 6 in. by 4½ in. by 6 in. long. The required curves are laid out on the face and adjoining edge of the block. The layout on the face is cut on a bandsaw with the laid-out face up. When the cutting is complete, the cutoff pieces are taped back into their original positions. These pieces provide support for the second cut. The block is rotated with the laid-out edge facing up. This is cut out. All the pieces are removed and the compound-sawed cabriole leg emerges. Smoothing and sanding produce a finished piece.

The drawing shows an alternative method for fabricating the front legs. A splined mitered joint-and-glue block is used in place of the glued-up block described above.

The rear legs are cut from two pieces joined at right angles, parts I and J. These are shown dovetailed together (side view of bottom rear of cabinet); however, a simpler joint can be substituted. In cutting this leg, care must be taken to ensure that a flat surface is always down on the bandsaw table during cutting. To attempt a cut on an unstable piece can result in saw blade fracture and possible injury to the operator.

The sides and top of the cabinet are made up of narrow pieces, maximum 4 in. width, joined edge to edge with dowels. Heartwood is reversed on each piece to help keep the pieces flat. These glued-up pieces can then be squared to finished dimensions.

The carcass sides have long dadoes cut into them with a straight router bit. The "drawer blades" or dust panel fronts fit into the dovetail portion of these dadoes. A detailed isometric blow-up drawing at the upper left side of the drawing shows how these dovetail dadoes are made. The first step involves cutting a regular dado across the carcass side using a router. In this case a T-square router base guide (shop made) is clamped to the carcass and guides the router. The router base is kept in contact with the T-square blade. All the dadoes are cut, seven in each side. Then a dovetail bit is placed in the router. In the second step, the first of each of the 2 in. sides of the original dado are beveled as shown. Since both sides are beveled, a dovetail-shaped recess is produced. This is done to the front 2 in. of the dado only. The rest of the dado is left in its original square shape.

The dust panels are made using frame-and-panel construction. (See the detail in the upper right exploded isometric view.) Note that the front of the dust panel, labeled the drawer blade, has a dovetailed-shaped end. The frame stile or side has a tongue-shaped edge. The dovetailed drawer blade end fits the first 2 in. of the dadoes made in the carcass side. This holds the carcass against the drawer blade and prevents bulging. The tongue-shaped edge of the stile (frame side) fits into the dado in the carcass side as it was initially cut with the router.

The drawer construction details are shown in the two views at the extreme left side of the drawing.

Note that the edges and ends of the top have the same shape as that of the ends and edges of the cabinet base. In addition, the drawer fronts are shaped with the same cutter that is used to shape the top and base of the cabinet.

Seven-Drawer Chest

Designed and Drawn by Raymond Langer March 1982.

Letter	Description	Number of pieces	Thickness	Width	Length
A	Drawer front	1	3/4	7	21
B	Drawer front	2	3/4	6	21
C	Drawer front	4	3/4	5	21
Da	Drawer side for drawer A	2	1/2	7	15
Db	Drawer side for drawer B	4	1/2	6	15
Dc	Drawer side for drawer C	8	1/2	5	15
Ea	Drawer back for drawer A	1	1/2	5-3/4	20-1/2
Eb	Drawer back for drawer B	2	1/2	4-3/4	20-1/2
Ec	Drawer back for drawer C	4	1/2	3-3/4	20-1/2
F	Drawer bottom	7	1/4	20-1/2	14-5/8
Ga	Drawer front hidden inside for drawer A	1	3/4	7	21
Gb	Drawer front hidden inside for drawer B	2	3/4	6	21
Gc	Drawer front hidden inside for drawer C	4	3/4	5	21
H	Front leg (two-piece)	4	1-1/2	4-1/2	6
I	Rear leg side	2	1-1/2	4-1/2	6
J	Rear leg back	2	3/4	4-1/2	6
K	Front drawer blade	8	3/4	2	22
L	Rear drawer blade	8	3/4	2	22
M	Dust panel	6	1/4	1-3/4	18
N	Side rails	16	3/4	2	11-3/4
O	Bottom of cabinet	1	3/4	15-1/2	24
P	Top of cabinet	1	3/4	15-1/2	24
Q	Back of cabinet	1	1/4	21-3/4	45-3/8
R	Sides	2	3/4	15	45

Notes
1. Materials: drawer fronts (hidden and exposed), sides, back, blades, and side rails, cherry; dust panels and back, luan plywood; top, bottom and sides, cherry or cherry-veneered plywood. Additional requirement: 14 drawer pulls.
2. Bottom and top side rails act as corner blocks screwed to bottom and sides with 1¼-in. No. 10 flathead wood screws.

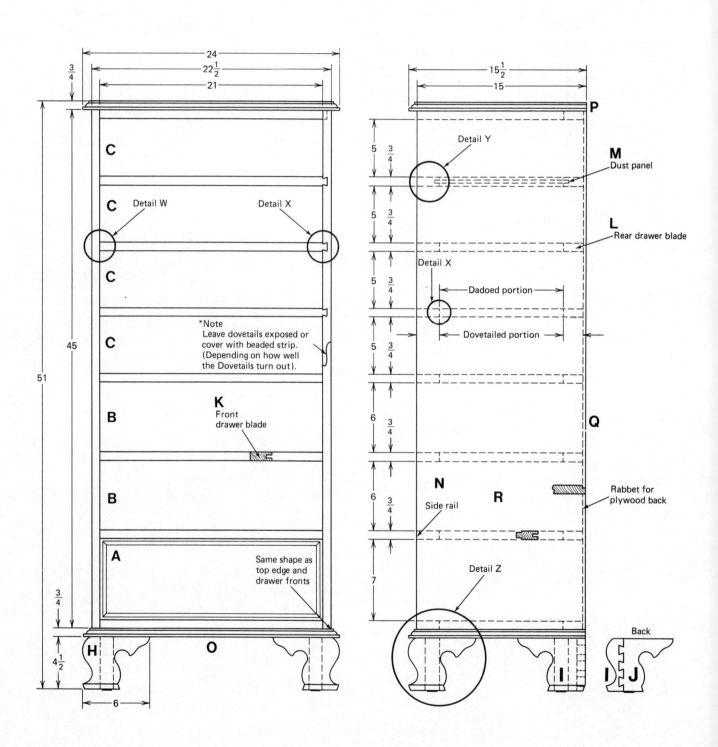

*Note
Leave dovetails exposed or
cover with beaded strip.
(Depending on how well
the Dovetails turn out).

K
Front
drawer blade

Same shape as
top edge and
drawer fronts

Detail Y

M
Dust panel

L
Rear drawer blade

Detail X

Dadoed portion

Dovetailed portion

Q

Rabbet for
plywood back

N
Side rail

R

Detail Z

Back

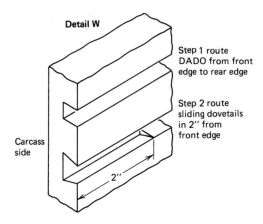

Detail W

Step 1 route DADO from front edge to rear edge

Step 2 route sliding dovetails in 2" from front edge

Carcass side

2"

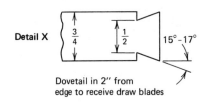

Detail X

$\frac{3}{4}$ $\frac{1}{2}$ 15°–17°

Dovetail in 2" from edge to receive draw blades

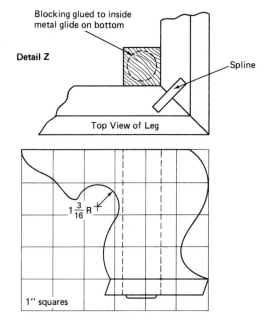

Blocking glued to inside metal glide on bottom

Detail Z

Spline

Top View of Leg

$1\frac{3}{16}$ R

1" squares

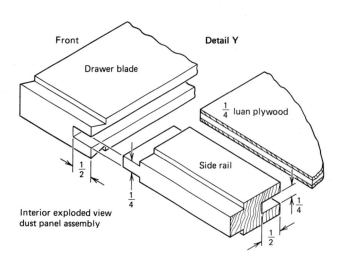

Front   **Detail Y**

Drawer blade

$\frac{1}{4}$ luan plywood

$\frac{1}{2}$

$\frac{1}{4}$

Side rail

$\frac{1}{4}$

$\frac{1}{2}$

Interior exploded view dust panel assembly

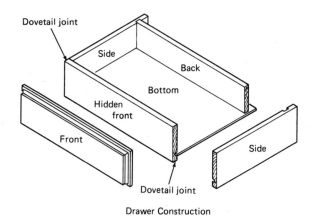

Dovetail joint

Side

Back

Bottom

Hidden front

Front

Side

Dovetail joint

Drawer Construction

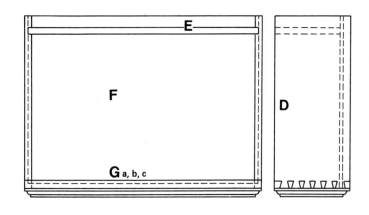

E

F

D

G a, b, c